THE GEOGRAPHY OF AIDS

About the Authors

Gary W. Shannon, a medical geographer, received his Ph.D. from the University of Michigan and is currently Professor of Geography at the University of Kentucky. His earlier work includes *Health Care Delivery: Spatial Perspectives* and he is the recipient of a Fulbright Western European Research Award to conduct an international comparison of physician relocation patterns in metropolitan regions. He is currently working on *Disease and Medical Care in the United States: A Medical Atlas of the Twentieth Century.*

Gerald F. Pyle is a Professor of Geography and Earth Sciences at the University of North Carolina at Charlotte. Educated at the University of Chicago, he has authored or co-authored a half-dozen books and numerous journal articles about the diffusion of disease. He has had several editorial responsibilities with *Social Science and Medicine* since the mid-1970s, and is active with the International Geographical Union Commission on Health and Development. He has lectured on medical geography in many countries, including Australia, Canada, the United Kingdom, Italy, and France.

Rashid Bashshur is Professor of Health Services Management and Policy at the School of Public Health, University of Michigan, and holds a Ph.D. from the University of Michigan. His teaching focuses on Health Services Research and Theory, as well as Health Resource Distribution and Allocation. His current research is concerned with development of health systems based on health need assessment in Alaska, as well as India, Pakistan, and Syria.

THE GEOGRAPHY OF AIDS
Origins and Course of an Epidemic

Gary W. Shannon
Gerald F. Pyle
Rashid L. Bashshur

THE GUILFORD PRESS
New York   London

© 1991 The Guilford Press
A Division of Guilford Publications, Inc.
72 Spring Street, New York, NY 10012

Printed in the United States of America

This book is printed on acid-free paper.

Last digit is print number: 9 8 7 6 5 4 3 2 1

Library of Congress Cataloging-in-Publication Data

Shannon, Gary William.
  The geography of AIDS : origins and course of an epidemic.

  Includes bibliographical references and
index.
    1. AIDS (Disease)—Epidemiology. 2. Medical
geography. I. Pyle, Gerald F. II. Bashshur,
Rashid, 1933–   . III. Title. [DNLM: 1. Acquired Immunodeficiency Syndrome. WD 308 S528g]
RA644.A25S43   1991   614.5'993   90-14112
ISBN 0-89862-445-2]

Preface

The complex course of the human immunodeficiency virus in its several forms appears to be moving inexorably, with new revelations about its characteristics and newly affected populations occurring almost monthly. Unless radical changes take place in the near future, either through development of a vaccine or a cure for those already infected, AIDS cases may total as many as 6 million by the end of the decade and up to 20 million people may be infected with the HIV. Social and public health scientists as well as the general public need to understand the complexities and scope of the disease and the related behaviors of its victims if they are to be successful in devising public and personal intervention strategies.

Our hope here is to provide the reader, regardless of specific interest or expertise, with a broader perspective and understanding of the current situation. In attempting to provide a timely and brief synthesis of such a dynamic and multifaceted subject certain constraints were inevitable. Each of the topics selected for discussion merits its own volume and some, for example, the immunology of the HIV, have been treated elsewhere in greater detail. Similarly, several edited volumes are available on selected aspects of the African experience. We make no claim nor was it our intent to treat exhaustively the topics selected for discussion. Thus, readers expert in immunology may be informed of the regional variation in the HIV experience but find our treatment of the HIV and its impact on the human immune system rather sketchy at the same time. Regional specialists in African, European, and American studies may quarrel with the limited treatment of their respective expertise but find discussion of the HIV immunology, competing theories regarding the origin of HIV, and the experience of other regions informative. Indeed, we have had to severely limit our discussions of the important and potentially explosive regions of Latin America and Asia, more from a lack of accessible information rather than design. This information is becoming available, however, and we hope to correct this situation in the near future. Finally, we are aware that our treatment of modelling the diffusion of the HIV will fall short in the estimation of theoretical geographers. We fully accept these shortcomings in our effort to provide a comprehensive yet timely, concise, and readable overview of the AIDS pandemic.

Lexington, KY   GWS
September 1990   GFP
                                                          RLB

Contents

1

An Overview

The acquired immune deficiency syndrome (AIDS) was discovered in the United States in the early 1980s, and by the spring
of 1990, about 140,000 cases of AIDS had been reported in this
country alone. This figure represents well over one-half of the
more than 200,000 cumulative cases of AIDS reported to the
World Health Organization (WHO, 1990) from 152 countries
around the world (Figure 1.1). And since AIDS cases are largely
underreported in the United States and in the rest of the world,
for logistic, social, and political reasons, the number of reported
AIDS cases represents only a fraction of the actual total. It is estimated that the actual cumulative number of AIDS cases to
date is at least double the number reported and, therefore, can
be reasonably expected to approach one-half million worldwide.

Compared with the millions of cases of other diseases nation-
wide and worldwide that existed in the same period, even one-
half million cases of AIDS over a 10-year period does not seem
particularly significant. When only the total numbers of AIDS
cases are considered and compared, there even appears to be
excessive concern with and attention directed towards AIDS. In
fact, as we become more accustomed or inured to the presence

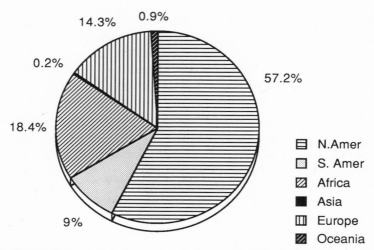

FIGURE 1.1. Relative world regional shares of actual AIDS cases reported to the World Health Organization by the beginning of 1990. Several
Third World countries, including Zaire and Thailand, began to report
substantial numbers of AIDS cases during 1990 (Source: World Health
Organization).

of the disease, arguments are being raised concerning the
amount of money allocated to researching the disease and to
developing vaccines or cures at the expense of the other major
killer diseases such as heart disease and cancer. Presently in the
United States, for example, an end to the special status of AIDS
studies at the National Institute of Health is being sought (Cordes, 1990). However, a closer look at AIDS quickly reveals that
the concern and attention are merited for several reasons.

First, the number of reported AIDS cases represents only the
"tip of the iceberg" with regard to the prevalence and incidence
of AIDS as well as the potential number of AIDS cases we can
expect in the near future. Underlying the numbers of actual
AIDS cases are the current and potential future numbers of persons infected with the purported etiological agent of AIDS, the
human immunodeficiency virus (HIV).

Admittedly, the situation is complex and dynamic with regard
to both current and forecasted numbers and patterns. The effec-

tive modes of HIV transmission appear to be limited, but the pattern, timing, and extent of HIV transmission depend heavily on individual behavior, social and cultural practices and customs, as well as biological factors related to individuals and to the virus. Short-term projections (over the next 4–5 years)of HIV infection and AIDS cases are relatively reliable since most of the projected AIDS cases will be developing in persons already infected with the HIV and are therefore not preventable. Because of the complexities and variable relationships of the factors related to HIV infection, long term projections (up to 10 years and longer) are much less reliable.

According to the WHO, through the end of the 1980s fewer than one million cases of AIDS had occurred throughout the world. In the United States current estimates of HIV prevalence vary between 800,000 and 1.2 million. However, some experts estimate that as many as 2.5 million persons are already infected in the Americas alone (Quinn, Zacarias, & St. John, 1989). According to the conservative WHO estimates, the number of persons infected with the HIV worldwide is about 6 million, but may in fact be over 10 million (Figure 1.2). By the year 2000, the WHO estimates that the number of HIV-infected adults may be as high as 20 million. Correspondingly, the total number of AIDS cases is also expected to increase substantially during the next 10 years, approaching 6 million by the end of the decade (Figure 1.3). And as information and estimates become available from countries which previously have withheld or purposely underreported HIV infection rates and AIDS cases, it is very likely this estimate will be revised upward significantly. Thus, today we find ourselves in the early stages of a global pandemic, the final stages of which can only be imagined.

Further fueling concern for the future of the pandemic are the facts that AIDS is a fatal disease and that there is currently, and in the foreseeable future, no vaccine against infection by the HIV, nor is there any cure available to those already infected. From a medical standpoint, therefore, the pandemic is coursing through the world's population, and there is little the medical and public-health communities can do to directly prevent the spread of the infection or cure those infected. By and large, the

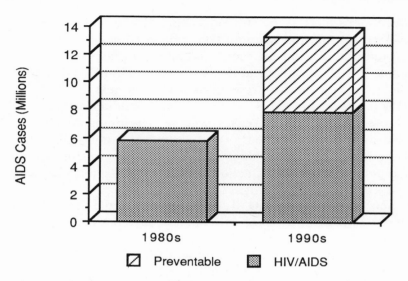

FIGURE 1.2 Projections of AIDS cases worldwide based on the Delphi technique wherein experts participate in "round robin" estimates until a general consensus is reached. Presumably, preventable cases during the 1990s are shown here by the hatched pattern (Source: World Health Organization).

medical community has been reduced to "damage control"—attempting to extend the lives and alleviate the suffering of persons infected with the HIV.

This is particularly unnerving to the medical and health care establishment in developed countries. Apart from the periodic influenza epidemic, it was generally believed that these nations had undergone an "epidemiologic transition." Medical advances had made infectious and contagious diseases—once ranked among the leading killers of populations—of little consequence. The largely chronic conditions such as heart disease, cerebrovascular diseases, and cancer had now become the leading killers, and even for these deadly diseases there was some chance for recovery and cure, to the extent that some can be reduced to chronicity.

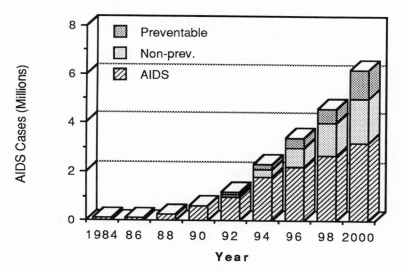

FIGURE 1.3. Delphi projections can also be used to show trends from 1984 to the year 2000. The number of preventable cases increases each year worldwide (Source: World Health Organization).

To be sure, in the United States and other developed nations recent improvements in life-expectancy are being documented for persons with certain AIDS-related diseases such as *Pneumocystis carinii* pneumonia (PCP), and one leading researcher recently suggested that by the end of the 1990s AIDS will be a "manageable" disease condition. Such optimism is apparently fueled by the improvement in life expectancy attributed to the application of certain recently developed and expensive drugs. At the same time, however, recent evidence indicated that life expectancy is decreasing for HIV- infected persons who develop an aggressive form of Kaposi's sarcoma (KS), a soft-tissue cancer. This cancer is now believed to be promoted by a protein produced by the HIV-1 (Ensolli, Barillari, Salahoddin, Gallo, & Wong-Staal, 1990). Until such time as effective vaccines or cures are developed *and* made available on a global scale, the only viable recourse to slow the pandemic is preventive educational programs directed toward changing human behavior

patterns and cultural traditions known to facilitate infection
with the HIV.

In the past, most educational campaigns have not met with
much success in changing health behavior. In this regard there
is some indication, though not yet definitive, that educational
programs directed toward safer sexual practices among male
homosexuals in the United States have contributed to at least a
reduction in the incidence of HIV infection among this group.
In the long run, however, if not already, this group may ultimately represent only a small fraction of the number of people
infected with the HIV. Globally, the majority of persons infected
with the HIV and developing AIDS presently and in the foreseeable future will be heterosexuals.

At present, Africa is the geographic area with the highest rate
of transmission of and infection with the HIV. Throughout the
developed world, the male-to-female ratio of persons infected
with the HIV ranges from 15:1 to 20:1. In most African nations
heavily affected by the infection, however, the male-to-female
ratio of infected persons is approximately 1:1. In certain age
groups, moreover, there is evidence that more females than
males may be infected. This same pattern is emerging in certain
Caribbean nations such as Haiti. At least in part, the developed
world's concern about the situation in Africa may be based on
the fear that it will also experience this pattern of HIV infection.
Thus, interest is directed toward understanding certain cultural
and health factors in African nations and determining similarities or differences in the AIDS situation there and in developed
nations. Clusters of heterosexual females infected by one promiscuous HIV-infected male have already been identified in Europe and the United States. Are male-to-female ratios of persons
infected with the HIV in certain African nations possible in the
developed world?

Apart from the issue of possible self-preservation on the part
of developed countries, there is genuine concern for the future
of developing countries in Africa and throughout the world.
Only recently have many countries in Asia and Latin America

experienced "seeding" of certain segments of their populations with the HIV. With large segments of the populations in these countries illiterate, the potential value of educational programs to prevent the spread of the HIV seems limited.

What are the implications, for example, of the rapid and widespread increase of HIV infection that is currently sweeping certain segments of Thailand's population (Anonymous, 1990)? The first case of AIDS was reported in that country in 1984; it was in a homosexual male who had just returned from a visit to the United States. By 1988, there were still only 10 cases reported, mostly among homosexual men believed to have contracted the virus during sexual liaisons while on foreign trips or with tourists visiting Thailand. By the end of 1989, however, the number of people infected with HIV had reached 13,600. By early 1990, the estimated number of HIV-infected persons is greater than 30,000 and expected to rise to over 1,000,000 by 1994. In Thailand it is now largely a heterosexual epidemic, spread mainly through heterosexual sexual intercourse in northern cities, and through intravenous drug use in the remainder of the country. In some northern cities over 70% of the "low class" prostitutes and 30 percent of the "high class" prostitutes have tested HIV positive.

Of equal concern must be recent reports indicating outbreaks of AIDS among intravenous drug users in China (WuDunn, 1990). In Yunnan Province on China's southwest border with Myanmar, formerly Burma, at least 146 peasants are reported to have tested HIV positive and an additional 100 blood samples from Yunnan peasants of Ruili Country heve been sent to Beijing for further testing. These peasants have been infected by sharing contaminated needles while injecting heroin.

Previously, AIDS as well as intravenous drug abuse was viewed by the Chinese government as a "foreign problem"(and the emphasis was placed on testing resident foreigners as well as short-stay tourists. According to officials, the increased use of heroin and the accompanying spread of the HIV in this corner of China is due to narcotics trafficking, as heroin is being transported through China to the West from the bordering countries of Myanmar, Thailand, and Vietnam. Drug addiction, however, is not new to China—an estimated 20,000,000 people were addicted to opium prior to the Communist revolution in 1949. Nevertheless, officials are adamant that China is not the source of the drugs, but is being used as a conduit because of its geographical location between drug producing countries and transportation "hubs" such as Hong Kong. In addition to the spread of the HIV through the sharing of contaminated needles among drug abusers it is also the consequence of the frequent reuse of hypodermic needles in medical settings and the relatively low rate of blood supply checks before transfusion.

From Rumania in eastern Europe emerges a picture of an AIDS tragedy affecting primarily infants and young children (Bohlen, 1990). Rumania is threatened with an unusual epidemic, the proportions of which remain unknown. Subsequent to the overthrow of the Communist dictatorship, health officials reported over 1,000 children found to test HIV positive. Most of the HIV positive and AIDS babies and children came from orphanages and group homes. They are believed to have accidentally received small amounts of HIV- contaminated blood, a "micro-transfusion," injected into their umbilical cord in the hope of stimulating growth. The practice of injecting blood from adults into infants and small children was practiced in the United States in the 1930s. At that time parent's blood was injected into children in a futile attempt to immunize them against measles. The practice is virtually unknown anywhere today and has not been practiced in the United States since the 1940s. In Rumania, it is believed that blood taken from a few HIV-infected adults was used for numerous small injections into children. In addition, some cases may be attributable to the re-use of contaminated needles resulting from a shortage of needles in the country (Hilts, 1990). while this type of pediatric AIDS problem is not believed to exist in other Eastern European countries, at least not on such a scale, it is certain that the number of reported AIDS cases from the German Democratic

Republic, Hungary, Poland, and Yugoslavia will rise well above the figures reported thus far. With the exception of Albania, AIDS cases have been reported from all Eastern European countries.

The dramatic spread of the HIV among Soviet children through repeated use of contaminated needles in Soviet hospitals accounts for much of the initial spread of the epidemic there. In early 1989 a needle used on an infected child spread the infection to 58 other children and nine adults in the town of Elista. The transfer of one patient to another hospital spread the infection to 27 additional persons (Bureau of Intelligence and Research, 1990).

In a speech to the Central Committee plenum in April 1989 Gorbachov warned of the potential for AIDS disaster. Although the Soviets had reported only 23 cases to the World Health Organization as of January 1990, their estimates indicate a much greater problem. While officially only about 500 Soviet citizens are infected, other estimates put the range between 1,000 and 10,000. By the year 2000, one official Soviet projection estimates the USSR may have as many as 15 million HIV-infected persons and 200,000 persons with AIDS (Bureau of Intelligence and Research, 1990).

The AIDS situation throughout much of the Middle East remains an enigma as tradition, culture, and religion reportedly have contributed to a "timid and tardy" recognition of the existence of the problem (Ibrahim, 1990). Some countries, such as Saudi Arabia and the United Arab Emirates, refuse to supply any information on AIDS cases, and other countries are believed to have grossly underreported the number of cases. In any event, the primary source of infection in the area appears to be contaminated blood and blood products imported into the area in the early and mid-1980s, before effective screening for blood banks and donors was established. Other sources of infection are reported to be HIV-infected prostitutes in the Sudan, Djibouti, and Somalia, as well as repeated use of contaminated needles among drug addicts. Drug addiction is believed to be on the increase in Egypt, Iran, Lebanon, and much of North Africa.

Participants in a recent conference said that Morocco, Tunisia, and Algeria had been exposed to the HIV epidemic primarily through their large numbers of migrant workers who live and work in Western Europe. There are an estimated four million Arab North Africans living in Western Europe, many in France. Only 322 AIDS cases have been reported to the World Health Organization from all Arab countries, plus Cyprus, Pakistan, and Somalia, a total of 23 countries.

In the Caribbean and Latin America, there are indications that some countries are headed toward an AIDS epidemic much like that affecting sub-Saharan Africa (Cortes, Detels, Aboulatia, et al., 1989). Of special concern is the emergence of a pattern of heterosexual spread of the infection. Though far behind the epidemic in Africa, there are signs the HIV infection will follow a similar course. Among the ominous indications are the relatively high proportion of female victims in some countries. For example, the male-to-female ratio in French Guina is 1.5:1; in Honduras, 1.7:1; in the Bahamas, 1.8:1; and, in Trinidad, 4:1.

In Honduras, the first AIDS case was diagnosed in 1985 in a homosexual dentist from Tegucigalpa who had visited San Francisco eight times in the two years prior to his diagnosis. By the end of 1988, 75 percent of the confirmed 182 AIDS cases were located around Tegucigalpa. However, 70 percent of the Honduran AIDS patients said they were heterosexual (Gruson, 1988). However, it is suggested that homosexuality is as prevalent in the Latin American region as in others, but it is not acknowledged because of the stigma associated with it in the traditional Latin "macho" culture, and discrimination against homosexuals may prevent homosexual AIDS victims from acknowledging that they are homosexuals. Also contributing to the rapid spread of infection in this region is the failure of many hospitals and clinics to test blood supplies for the HIV antibodies and the high prevalence of genital infections and sores.

It is apparent from the preceding discussion that for many world regions and the millions of people living in them there is presently scant knowledge concerning the status of the HIV epidemic and numbers of AIDS cases. There is sufficient information, nevertheless, to cause great concern in these regions as the HIV begins to move into various populations initially susceptible to the virus through either particular behaviors or medical practices and conditions. To date, the people and countries of sub-Saharan Africa appear to be at greatest risk for the potential destructive force of the pandemic. As we move into the 21st century, however, will we witness HIV infection rates in other countries and regions similar to the pattern emerging n certain African nations? If so, what will be the regional and international demographic, economic and political consequences?

The HIV and AIDS are novel and complex phenomena. Understanding them requires the concentrated and cooperative efforts of both the medical–biological and social sciences. Hopefully, the ultimate answer will come from medical–biological research in the form of vaccines to prevent infection and drugs to rid the body of the virus. Meanwhile, it is imperative for social scientists to concentrate their efforts on understanding the virus and its modes of transmission. Based on this understanding, they should examine the social–behavioral context of HIV infection and AIDS, and should aid in the development of programs appropriate to controlling the infection process and caring for those infected.

In this volume we attempt to describe and illustrate some of the spatial and temporal dimensions of the pandemic using available data. In the course of our discussion, the complex interplay of seemingly myriad cultural, social, psychological, political, and economic factors becomes evident. We contend that an understanding of these factors must be based on the concept of "region." Only by focusing on regions can the relevant factors be identified and the interplay of these factors sufficiently understood to contribute to manageable and meaningful discussion.

We begin this volume by describing the structure of the HIV, its life-cycle process, mode of transmission, and impact on the human immune system and other parts of the body. From there we proceed to a discussion of the major theories pertaining to the geographic origin of the HIV. Of the regions to be considered, Central Africa is the first, since it appears to be especially important as a possible index location for the HIV as well as the potential for human and economic disaster. Chapters on the progression of AIDS in Europe and in the United States follow. We then focus on the implications of AIDS for health care systems generally and particularly those of the United States. Following these discussions, we focus on models that replicate and forecast the spatial and temporal diffusion of HIV infection and suggest revisions thereof.

2

"A Strange Virus . . ."

The title of this chapter derives from a book written 5 years ago dealing with the human immunodeficiency virus (Leibowitch, 1985). The appellation was perhaps more accurate at that time. In fact, despite allegations of initial confusion and lack of attention by federal public health officials and the medical research community, progress in understanding the HIV and the etiology of AIDS has been relatively rapid (Cahill, 1984; Baltimore & Feinberg, 1989; Ho, Moudgil, & Alam, 1989).

The actual discovery of the HIV remains the center of an international dispute between researchers in Paris and Bethesda, Maryland (Crewdson, 1990). In 1983 researchers at the Pasteur Institute in Paris isolated a novel human retrovirus from a lymph node biopsy of a Parisian fashion designer, a male homosexual with lymphadenopathy (swollen lymph nodes) and labeled the agent the lymphadenopathy virus (LAV) (Gallo & Montagnier, 1988). Some months later at the National Institutes of Health in Bethesda researchers also isolated the same type of virus and further determined that it was "attracted to" or had a selective tropism for a cell critical to the human immune response system—the T4 or T4- helper/inducer cell (Fauci, 1988). Thus, they labeled the virus the human T-Cell lymphotropic virus type III (HTLV-III). Apparently there is an increasing amount of evidence to indicate that the NIH-isolated virus was identical to and cultured from the virus isolated by the researchers of the Pasteur Institute. In any event, the numerical designation reflected the relationships between the virus and two other human retroviruses identified in the late 1970s: the first associated with adult T-cell lymphocytic leukemia and rare tropical neurologic illnesses, and the second isolated from patients with a variant of hairy cell leukemia (Kalyanaraman, Sarngadharan, & Gorodff, 1982; Weiss & Biggar, 1986).

In the short time since its discovery, much has been learned and hundreds of scientific articles have revealed details about this once "strange virus." Before proceeding with the discussion of the geography of AIDS, we feel it is important to present some basic information about the virus believed to be responsible for AIDS, including its family'background, structure, life cycle, infectious process, and mode of transmission.

HIV AS THE DISEASE AGENT

Because of early confusion and debate pertaining to the newly recognized AIDS-related virus and its properties, in early literature the virus was variously labeled: (a) lymphadenopathy associated virus (LAV); (b) human T-Cell lymphotropic virus type III (HTLV-III); and (c) AIDS-related virus (ARV). Only relatively recently has the international research community agreed on the name human immunodeficiency virus (Adler, 1987; Centers for Disease Control, 1987a, 1987b). With the isolation of a related variant of HIV in West African AIDS patients now labeled HIV-II or HIV-2, the "original" virus is now labeled HIV-I or HIV-1 (Clavel, Mansinho, & Chamanet, 1987; Denis, Gershy-Damet, et al., 1987; Mabey, Tedder, et al., 1988).

The overwhelming evidence points to the HIV(s) as the specific etiologic agent(s) for AIDS. Epidemiologic proof of the HIV as the cause of AIDS generally follows revised guidelines based on Koch's original postulates. These very conservative and restrictive axioms serve as the basis for establishing a microorganism as the etiologic cause of a disease. As originally set forward, the disease agent must (1) be found in all cases of the disease, (2) be isolated from the host and grown in pure culture, (3) reproduce the original disease when introduced into a susceptible host, and (4) be present in the experimental host so infected.

Opponents of the HIV-1 as the AIDS disease agent (Duesberg, 1987, 1989) appear to base their arguments on the most strict interpretation of Koch's postulates, disregarding the state of medical knowledge and technology regarding the complexity and capabilities of the virus as well as our ability to locate the HIV-1. For example, some early studies demonstrated a low rate of isolation of the HIV-1 in infected persons (Salahuddin, Markham, Popovic, et al., 1985). However, recent quantitation of the HIV-1 in the blood of infected persons resulted in much higher levels of infectious HIV-1 than previously estimated and a significant increase in the HIV-1 in persons at various stages of infection—from an asymptomatic stage to AIDS-related complex (ARC) conditions to, finally, full-blown AIDS (Ho et al., 1989). No infectious HIV-1 was found in any of the seronegative controls. For almost any disease, it is only in the rarest instances, if ever, that Koch's conditions to establish causation can be fully and universally achieved. Certainly, at any given time pathogens may be present but remain undetected, given contemporary levels of medical knowledge and technology. Evidence suggests the HIV-1 is capable of "hiding" from the immune system; hence the current tests to detect it. Some individuals appear to be persistently infected, but the virus is not replicating at a level sufficient to induce antibody formation. This condition can exist for years (Moss & Bacchetti, 1989). Therefore, epidemiologists are forced to rely on such measures as biological plausibility, consistency, relative risk, biological gradients and coherence, and statistical associations to determine the pathological cause of many diseases. The relationship between the HIV-1 and AIDS is no exception.

Researchers accordingly designated the HIV-1 as the cause of AIDS because of (1) its repeated isolation from AIDS patients and persons at high risk of AIDS; (2) numerous cross-sectional studies indicating significantly more antibodies against the HIV-1 in AIDS patients than controls; (3) the spatial and temporal covariation in the proportions of risk-group members testing seropositive for the HIV-1 coincident with the spatial and temporal trends of the epidemic; and (4) the absence of HIV-1 antibodies in historic sera obtained prior to the onset of the AIDS epidemic, even in high-risk-group members and persons with a variety of clinical illnesses considered to be suggestive of AIDS (Weiss & Biggar, 1986). Two major observations appeared to solidify the causal relationship. First was the repeated finding, in almost all instances, of a seropositive donor in a study of blood and blood product transfusion-associated AIDS cases. Second, in prospective studies of homosexual men and parenteral drug abusers, only those with preexisting exposure, as demonstreted by the presence of HIV-1 antibodies, went on to develop the immunodeficiencies and clinical syndromes characteristic of AIDS (Weiss & Biggar, 1986).

THE HIV FAMILY

Basically, like all viruses, the HIV is a sub-microscopic intercellular parasite: the virus particle itself is inert and cannot propagate or do any damage until it enters a host cell (see Figure 2.1). The portion of the virus essential to its survival is ribonucleic acid (RNA) which contains the information necessary to its functioning and reproduction. It is endowed with chemicals and properties that permit it to be reproduced (Leibowitch, 1985). However, to replicate itself, the virus must "borrow" the necessary ingredients from other cells composed of deoxyribonucleic acid (DNA). The retroviruses are capable of integrating into the chromosome of a host cell and reversing the viral nucleic acid to cellular nucleic acid. The retroviruses possess the chemical equipment, including the enzyme reverse transcriptase, necessary to copy themselves before "returning" into the chromosome of the host cell.

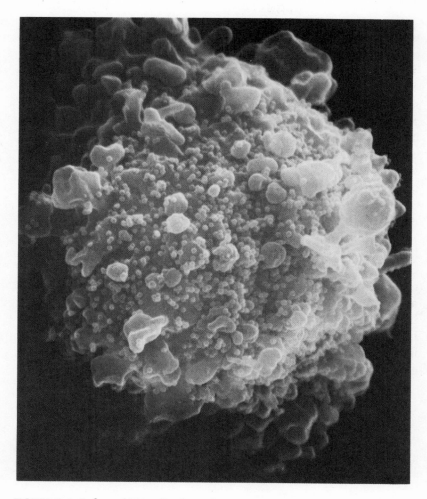

FIGURE 2.1. Infected T4 cell. The infected T4 cell produces HIV virions (small spheres) of the HIV in this image. The scanning electron micrograph shows part of the infected cell's convoluted surface (magnification: ~× 12,800). (Photograph courtesy of David Hockley of the National Institute for Biological Standards and Control, Hertfordshire, England.)

The retroviruses were so named because of what appears to be a process of genetic expression that is the reverse of the usual genetic expression (Haseltine & Wong-Staal, 1988). The genetic basis of cells is DNA, and when genes are expressed, the DNA is first transcribed into "messenger" RNA, which then serves as the "template" for the production of proteins. In contrast, the genes of the retrovirus are encoded in RNA and, before they can be expressed, the RNA must be converted into DNA. Only upon completion of this reverse process can the viral genes of the retrovirus be described and translated into protein in the usual sequence (Haseltine & Wong-Staal, 1988).

The HIVs belong to the lentivirus subfamily of human retroviruses (see Table 2.1) (Montagnier & Alizon, 1987; Levy, 1988, 1989). This family is distinguishable from the other subfamilies of human retroviruses, the Spumavirinae and Oncovirinae on the basis of morphologic and morphogenetic characteristics (Levy, 1989). Specifically, the lentivirus genomes are large and contain several viral genes. The viruses frequently induce cytopathic effects in infected cells; the disease they cause has a long incubation period resulting in slowly progressive immunologic disorders and neurological disease, which are inevitably fatal (Fauci, 1988).

TABLE 2.1. Human Retrovirus Subfamilies[a]

Lentivirinae (nontumorigenic)
    HIV-1
    HIV-2

Spumavirinae
    Human foamy virus

Oncovirinae (tumorigenic)
    HTLV-I
    HTLV- II

[a]Adapted from Levy (1989) and Najera, Herrera, and Andres (1987).

HIV-1, HIV-2, AND NONHUMAN PRIMATES

When the genetic sequence of HIV-2 was reported, it was clearly significantly different from HIV-1, as indicated in Table 2.1, although the two viruses were nevertheless more closely related to each other than to any other retroviruses whose sequences were known at the time (Doolittle, 1989). Although retroviruses have been found in other animal species, including sheep and goats (maedi-visna virus), horses (equine infectious anemia virus), and cats (feline immunodeficiency virus), the genetic sequence of these viruses indicates that they are only very distantly related to the HIV-1 and the HIV- 2.

The HIV-1 and the HIV-2 have a similar structure, although identifiable genes within each subtype can be found. Moreover, there are apparent biological and serological differences between the two subtypes in terms of cellular host range, extent and kinetics of virus replication, cytopathology, induction of latency, and modulation of CD4+ antigen expression on infected cells (Levy, 1989). Both cytopathic and noncytopathic HIV-2 strains have been identified.

The HIV-2 subtype was initially found only in West Africa and remains endemic in this part of the world, but infected individuals have been reported in other parts of Africa, Europe, and South America. In some instances persons have tested positive for both HIV-1 and HIV-2. This may be due to either true dual infection or to infection with one virus that has cross-reactivity with the other. The discrimination of infection is of increasing serologic importance since both viruses seem to have the same modes of transmission and HIV-2 may be spreading into areas in which the HIV-1 is already endemic (Quinn, Zacarias, & St. John, 1989). The first documentation of an HIV-2 AIDS-related infection in the United States was in a female AIDS patient who had immigrated to New Jersey from West Africa after the onset of her illness in 1987 (Centers for Disease Control, 1988a). A second case has now been identified, a 34-year-old Caucasian woman who was born and lived in West Africa until 1979 when she divorced her West African husband and immigrated to the United States (Ruef, Drekey, Schable, 1989). In 1981 she remarried a U.S. resident, who was also originally from West Africa (and subsequently tested negative for the HIV). She was admitted to a hospital in May 1988, 9 years after coming to the United States. Although indigenous HIV-2 infections have been documented in France (Brucker, Brun-Vezinet, Rosenheim, et al., 1987; Dufoort, Courouce, Ancelle-Park, & Bletry, 1988), none have been detected to date in the United States.

Since 1984 several nonhuman primate lentiviruses—the simian immunodeficiency viruses (SIVs)—have been identified. As we shall see in subsequent chapters, the identification of these SIVs is central to hypotheses pertaining to the biological and geographic origin of the HIVs. Attention centers on nonhuman primates as a possible source of cross-infecting viruses (Doolittle, 1989). Within a short period after isolation of the HIV-2 in humans, the sequence of an SIV (SIVmac) isolated from a captive macaque monkey suffering from an AIDS-like illness was determined to be remarkably like that of the HIV-2. With few exceptions, however, macaque monkeys are indigenous to Asia and not Central Africa. Further, it is believed that wild macaques do not harbor the SIVmac. It is likely that the captive macaque was infected during captivity, probably by mangabey monkeys indigenous to West Africa. In fact, a recent study found that a molecularly cloned SIV of the sooty mangabey monkey (SIVsm) and the HIV-2 are quite closely related (Hirsch, Olmsted, Murphey-Corb, et al., 1989).

On the other hand, the HIV-1 and the HIV-2 appear to be equally distant from the SIV isolated from the African green monkey (SIVagm) found throughout Central Africa. The search continues for more closely related SIVs from an as yet unidentified nonhuman primate. Against this background, the debate over the simian–human connection remains vigorous in some corners (Seale, 1988, 1989). Another line of investigation focuses on the relatively rapid and uncontrolled genetic drift of the HIVs themselves to explain the possibility of species transfer.

HIV-1 BASIC STRUCTURE

The retrovirus is constructed of RNA rather than DNA, unlike the typical virus and cell (Gelderbloom, Hausman, et al., 1987; Weber & Weiss, 1988). As illustrated in Figure 2.2, the HIV-1 has a dense cylindrical, protein central core encasing two molecules of the viral RNA. The core shell consists of p24. The core protein, p17, is found outside the viral nucleoid and forms the matrix of the virion, while p15 apparently represents another basic core protein. Located on each of the RNA molecules are the enzymes reverse transcriptase (RT) and integrase. Again, these substances enable the transcription of the viral RNA genome into a DNA copy that eventually integrates into the

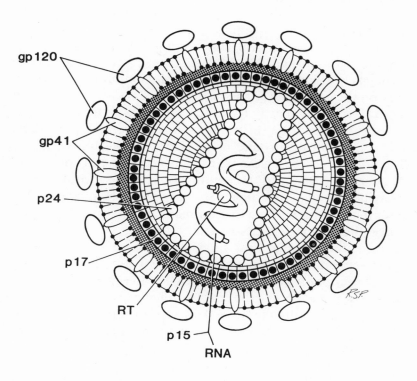

FIGURE 2.2. HIV-1 structure in cross section.

host cell chromosomal DNA (Levy, 1989; Haseltine & Wong-Staal, 1988; Gallo & Montagnier, 1988; Klatzmann & Gluckman, 1987). As discussed below, also of particular importance to the life cycle of the HIV and the infection process are the glycoproteins located on the envelope of the virion. The major envelope glycoprotein (gp120) is located on external "spikes" connected to the virion by the transmembrane glycoprotein (gp41).

LIFE CYCLE AND THE INFECTIOUS PROCESS

The initial phase of viral reproduction occurs at the cell membrane (see Figure 2.3). There appear to be certain proteins on the cellular surface that serve as receptors for the glycoproteins on the surface of the virion. In particular, the CD4 protein has been identified as an efficient receptor for the viral envelope glycoprotein (also gp120). Of particular importance also is the viral gp41, which interacts with some fusion protein (F) receptors on the cell surface. The presence of CD4 or F receptors of several types (such as FcRI, FcRII, and FcRIII) individually on a cell surface may be sufficient for initial interaction or "docking" of the virion to take place (Levy, 1989; Homsy, Meyer, Tateno, et al., 1989). Cells whose surfaces contain some combination of CD4 and F receptors would make a viral docking more likely and more efficient. It may be that the fusion of gp41 with the cellular surface membrane is a prerequisiue for viral entry (Stein, Gowder, Lifson, et al., 1987).

Once docked, the viral nucleoid or core enters the cell cytoplasm and the reverse transcriptase within the core begins the transcription process. The RNA/DNA copy produced then merges with the DNA of the cell. The infected cell is now capable of simply harboring the RNA/DNA copy or reproducing copies of the retrovirus at the cell surface through a process known as budding.

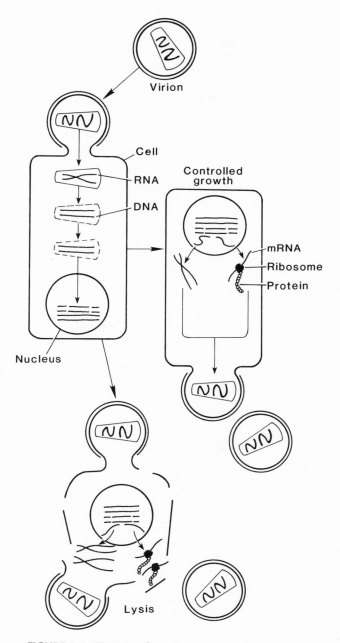

FIGURE 2.3. HIV-1 infection and reproduction.

INFECTION SITES

The multiple docking sites provided by the protein receptors of various types allow for numerous types of cells to be infected with the HIV. To date, although it has only recently been established that the HIV can efficiently infect a cell independently of the CD4 protein, the distribution of cells infected closely reflects the distribution of cells bearing the CD4 molecule. In body tissues, for example, with the exception of glial cells in the brain and chromaffin cells in the colon, duodenum, and rectum, infected cells generally carry the CD4 molecule on their surface (see Figure 2.4). The HIV has also been identified in or isolated from a number of bodily fluids (see Table 2.2). Of particular importance are those to which individuals are most likely to be intimately and directly exposed such as semen, blood and blood products.

Especially important to the development of AIDS is the infection of various cells of the cellular human immune response system including the T4-helper/inducer lymphocytes, B-lymphocytes, and monocytes/macrophages. Of particular importance is the infection of T4-helper cells. These cells have been labeled the orchestra leader of the human cellular immune response system because of their direct and indirect responsibility for a wide array of functions (see Figure 2.5). The T4-helper cells multiply and produce lymphokines which in turn regulate B- and T-cell production. In addition, interaction between macrophages and T4-helper cells stimulate T8-"killer" cells to mature and destroy infected cells. Simultaneously, the receptors on the B-cells also receive signals from the lymphokines, proliferate and secrete antibodies that neutralize invading viruses directly. The T4-helper, being a primary target for the HIV, thus has important implications for the wholesale impairment and destruction of the cellular immune response system. To compound the problem, the CD4 receptor has been found not only on T4-helper cells, but B-cells, macrophages, and some brain cells as well (Kolata, 1988; Levy, 1988).

Once infected, the integration of the viral RNA with the cellular DNA takes place. In the process the HIV may have a

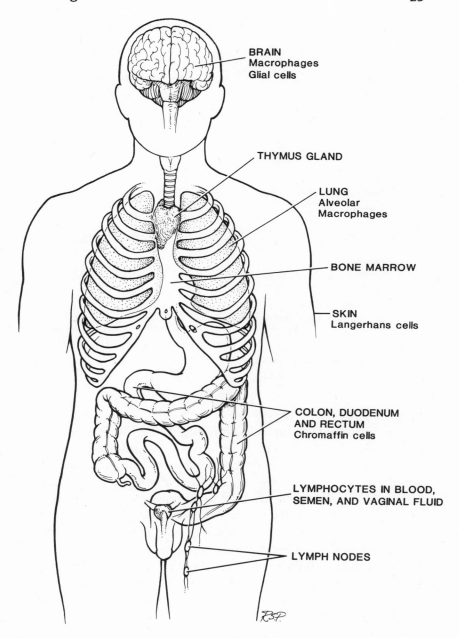

FIGURE 2.4. Distribution of HIV infection sites.

TABLE 2.2. Distribution of the HIV in Human Body Fluids

HIV isolated from:	Quantity (HIV/ml)
Cerebrospinal fluid	10-1000
Plasma	10-50
Semen	10-50
Serum	10-50
Ear secretions	5-10
Tears	<1
Saliva	<1
Urine	<1
Vaginal/cervical fluids	<1
Breast milk	<1
HIV identified in infected cells:	**Percent Infected**
Semen	0.01-5
Peripheral blood/mononuclear cells	0.001-0.1
Saliva	<0.01
Bronchial fluid	*
Vaginal/cervical secretions	*

* = Unknown
Note: Adapted from Levy (1989).

cytopathic effect on the cell, resulting in the diminished function or death of the cell. Alternatively, the infected cell may
become a "factory" for the production of copies of the HIV
which, in turn, infect other cells. The capacity of the immune
response system can be affected in several ways. The role of the
T4-helper/inducer cells can be reduced in its ability to orches-
trate the other segments of the immune system, or they can be
destroyed outright. Alternatively, they may also serve as factories for the production of "copy" HIVs which then infect other
T4-helper/inducer cells as well as other cells important to the
immune response system. In the course of this reproductive
phase the host cell is usually killed, possibly through a massive

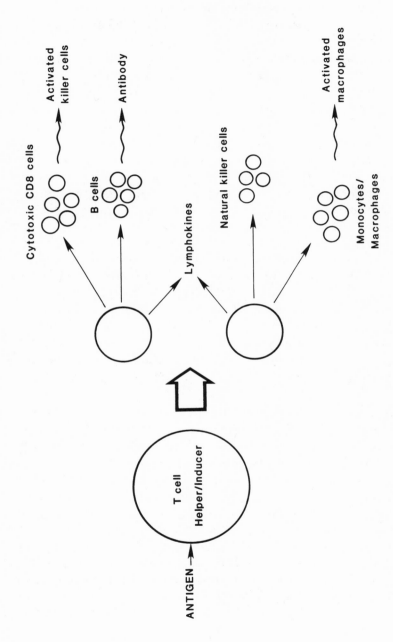

FIGURE 2.5. T4-helper/inducer cell functions.

increase in the permeability of its cellular membrane in the budding process or in the accumulation of unintegrated viral DNA. In addition to the direct negative impact on the number and capacity of T4-helper/inducer cells, which reduces the immune system's ability to stimulate the production of other pathogen-fighting cells, the capacity of other cells with important immune responsibilities including the B-cells and macrophages can be destroyed or significantly reduced through direct HIV infection. Since the level of expression of CD4 on these monocytes is less than on the T4-helper/inducer cell, the impact of the HIV is not as great. In any event, persons with AIDS suffer an enormous disruption of their immune system with severe depression of numbers and capability of specific cells critical to the lymph cell-mediated immunity.

It should be noted here that recent studies using newer and more sensitive detection techniques have identified much higher levels of the HIV-1 in AIDS patients and patients with AIDS related conditions (ARC) such as high fever, lymphadenopathy, and weight loss. Further, significant increases in HIV-infected lymphocytes have been identified as ARC and AIDS progresses (Ho et al., 1989). Another study found that, once detected, the HIV is always found in the blood, thus challenging the notion that it may be dormant at times. Further, the presence of the HIV in the blood was more closely correlated with the clinical stage of disease than results of other tests (Coombs, Collier, Allain, et al., 1989). The possibility is raised, therefore, that the direct cytopathic effect of the retrovirus alone may be sufficient to explain much of the pathogenesis of AIDS.

OPPORTUNISTIC DISEASES

In any event, with a severely compromised immune system, an individual is susceptible to a host of infections and diseases that are not threats to individuals whose immune systems have not been depleted by the effects of HIV infection. Hence, these diseases and conditions are termed *opportunistic* because they in effect take the opportunity to develop against a depleted immune system. A number of diseases are associated with cellular immune deficiency (see Table 2.3).

Initial publications described two major opportunistic diseases ultimately associated with AIDS: Kaposi's sarcoma (KS), a soft-tissue cancer extremely rare in the United States or

TABLE 2.3: Infections and Disease conditions Indicative of Cellular Immune Deficiency

Protozoal and helminthic infections:
    Cryptosporidiosis (chronic enteritis)
    Isosporiasis
    Pneumocystis carinii (pneumonia)[a]
    Stronglyoidosis (pulmonary, central nervous system [CNS], or disseminated)
    Toxoplasmosis (encephalitis or disseminated)

Fungal Infections:
    Aspergillosis (CNS or disseminated)
    Candidiasis (esophageal or broncho-pulmonary)
    Cryptococcosis (pulmonary, CNS, or disseminated)
    Histoplasmosis (disseminated)

Bacterial infections:
    "Atypical" mycobacteriosis (species other than tuberculosis or lepral; disseminated)

Viral Infections:
    Herpes simplex virus (chronic mucocutaneous or disseminated)
    Cytomegalovirus (pulmonary, gastrointestinal, or CNS)
    Papovirus (progressive multifocal leukoencephalopathy)

Cancers:
    Kaposi's sarcoma (age <60)
    Cerebral lymphoma
    Non-Hodgkin's lymphoma (diffuse, undifferentiated)
    Lymphoreticular malignancy

Others:
    Chronic lymphoid interstitial pneumonitis (age <13)

[a]Recently identified as a fungus infection (Edman, 1988).
Sources: Adler (1987) and Peterman, Drotman, & Curran (1985).

among young persons, and an equally rare form of pneumonia, *Pneumocystis carinii* pneumonia (PCP), among small groups of homosexual males in California and New York in 1981 (Centers for Disease Control, 1981a, 1981b). In developed societies both KS and PCP as well are among the most frequent life-threatening health problems for persons with AIDS (Armstrong, Gold, et al., 1985; Biggar, 1986). In less developed countries Cryptosporidiosis or chronic enteritis and associated cachexia (wasting), toxoplasmosis, and tuberculosis are most frequent. Other unexplained international variations in opportunistic diseases associated with AIDS have been found in the occurrence of PCP. While most frequent among patients in the United States, PCP was rarely found among the early AIDS patients in the United Kingdom (Weber, Carmichael, et al., 1984).

The clinical expression of infection with the HIV appears increasingly complex. It includes manifestations due to opportunistic diseases, as well as illness directly caused by the HIV itself. For example, the geographical distribution of tuberculosis as an opportunistic disease in AIDS patients varies considerably. In most instances it is viewed as the reactivation of a previous tubercular infection (Armstrong et al., 1985; Sunderam, McDonald, et al., 1986). Therefore, the distribution of AIDS patients developing tuberculosis is believed to be largely coincident with the distribution of populations at risk for tuberculosis reinfection.

In the United States KS appears much more frequently among AIDS patients who are homosexual males than parenteral drug abusers. Also, KS is seen more frequently in AIDS patients from New York and California than elsewhere in the country. This prompted researchers to look for an additional behavioral cofactor that might explain this spatial variation. Some suggested that behavioral factors, such as oral-fecal contact and "recreational use" of non-labeled amylnitritie inhalants, appear to predispose an individual to KS (McKusick, Conant, & Coats, 1985; Safai, Johnson, et al., 1985). More recent speculation centered on a novel growth factor released by HIV-1-infected cells being instrumental in developing the aggressive form of KS seen in AIDS patients (Newmark, 1988). However, a recently published report again suggests that there may be a cofactor operating independently of the HIV infection that causes KS among male homosexuals (Beral, Peterman, Berkelman, et al., 1990). It is probable that the types of infections and neoplasms seen in persons with AIDS may vary not only in populations of different geographic origin and susceptibility but also according to the way the HIV infection was acquired (Piot & Colebunders, 1987).

HIV TRANSMISSION

The HIV-1 has been isolated from many tissues, organs, and fluids of infected people, including peripheral blood lymphocytes, lymph nodes, bone marrow, brain and kidney tissue, blood cell-free plasma, and in free-form or infected cells of saliva, semen, vaginal/cervical fluids, cerebrospinal fluid, and breast milk (Levy, 1989; Vogt, Witt & Croven, 1986; Wkofsy, Cohen, & Haver, 1986; Gallo & Sarngadharan, et al., 1987). This near omnipresence does not mean the tissues and particularly fluids transmit the infection equally since their HIV-1 concentration as well as the likelihood and type of exposure varies considerably. Semen, blood, and vaginal/cervical fluids and secretions are particularly important sources of infection, depending upon behavior and host susceptibility (Adler, 1987; Levy, 1988). The transmission of the HIV takes place through intimate sexual contact, contaminated blood and blood products, and passage from mother to child during perinatal events (Levy, 1989). Epidemiological and immunological studies in Europe, the Americas, Africa, and Australia (Adler, 1987; Winkelstein, Lyuman, et al., 1987; World Health Organization, 1987c; Castro, Lieb, et al., 1988; Jaffe & Lifson, 1988; Lifson, 1988; Levy, 1989) repeatedly document three major modes of HIV-1 transmission:

1. Sexual (heterosexual or homosexual) intercourse (the most common mode of transmission);
2. Contact with blood, blood products, or donated organs

and semen (the vast majority of contacts with blood in-
volve transfusion of unscreened blood or the use of un-
sterilized syringes and needles by parenteral drug abusers
or in other settings); and,

3. Mother to child—mostly before and perhaps during or
   shortly after birth (perinatal transmission). Possible transmission via breast milk appears to be rare as only two
   cases of this possible transmission mode have been recorded.

SUMMARY AND CONCLUSION

Since the initial isolation of this "strange virus," much has been
learned about the structure, life cycle, and impact of the HIV,
but much remains to be learned. The HIV-1 and HIV-2 represent "novel" retroviruses which are relatives, albeit distant, of
retroviruses found in some other nonhuman animals. The HIV-
1 is clearly different from the HIV-2, although both are more
closely related to each other than to any other known retroviruses whose genetic sequences were known at the time of
discovery. The HIV-2, especially, is closely related to SIVs recently discovered in various species of African monkeys such as
the African green monkeys, mandrills, and especially the
sooty mangabeys.

Both HIVs have been isolated from AIDS patients, but to date
the HIV-1 is most frequently associated with the severe com-
promise of the human cellular immune system and the resulting expression of AIDS. Recent evidence suggests, alternatively,
that the HIV is never dormant and that the amount of the HIV
present may be sufficient to explain most of the damage that the
body suffers from AIDS. The HIV-2 appears to be less cytopathic than the HIV-1 and appears to have a longer period of
latency or alternatively reproduces more slowly than the HIV-1.
Apparently important to infection by the HIV is the presence of
certain glycoproteins on the surface of many different types of
cells in the human body which makes them "attractive" to the
HIV. While many parts of the body may be directly affected by

the infection, the primary impact of the HIV is on the cellular immune system. There now appears to be some evidence of the presence of a microbe, called a mycoplasma, which may play a key role in the destruction of the human cells by the HIV (Lemaître, Guétard, Henin, Montagnier & Zerial, 1990; Wright, 1990). The mycoplasma, a primitive microorganism that lacks a cell wall, may be an important cofactor in HIV infection and associated with infections that speed the progression to full-blown AIDS or cause the occurrence of opportunistic infections late in the course of the disease.

The primary mode of HIV transmission is through intimate contact with infected blood and blood products, semen and cervical/vaginal fluids, or perinatal events. In developed countries those people particularly at risk to infection have been homosexual males, hemophiliacs, parenteral drug abusers, and children of infected mothers. In underdeveloped countries, especially in Africa and the Caribbean, the infection appears to spread predominately by heterosexual intercourse. To date, there is no vaccine to prevent infection and no cure for those people infected. Prevention can be accomplished by behavioral change.

A number of theories have been advanced regarding the geographic origin of the HIV. As well, there are significant differences in the geographic distribution of AIDS and the underlying modes of transmission. Dramatic geographic differences in HIV infection and related AIDS cases have been demonstrated. In succeeding chapters we attempt to elaborate on selected geographic aspects of the HIV and AIDS.

3

" . . . Of Unknown Origin"

According to the astrophysicist Sir Fred Hoyle, the human immunodeficiency virus is of extraterrestrial origin (McClure & Schulz, 1989). He and his colleague have also theorized that influenza virus molecules floating through space might be driven into the earth's atmosphere by solar winds created during peak sunspot activity (Hoyle & Wickramasinghe, 1990). It has been suggested that the virus was created artificially either deliberately as a biological warfare weapon by the "doctors of death" at Fort Deitrich, Maryland, or accidentally by molecular biologists in a recombinant research laboratory of some sort in the U.S.S.R. or Eastern Europe (Anonymous, 1988a). One hypothesis suggests it is plausible to suspect that the AIDS epidemic was started deliberately in the United States "by a hostile power" (Seale, 1988).

A hypothesis of a similar genre links the HIV to an African strain of swine virus causing deadly fever among Cuban hogs, the source of the virus being either importation of infected stock or the direct and malicious contamination of the Cuban hogs, financed by the Central Intelligence Agency (Leibowitch, 1985). Others maintain that perhaps it was not venereal syphilis at all but the HIV infection that Columbus's sailors brought back from Hispaniola (Haiti) in 1493 (Andre, 1987). A Euro-American origin has been traced based on the similarity of the HIV to a visna-maedi virus found in northern European sheep and the transfer of the virus to humans implied through sexual contact between human males and sheep (Kanter & Pankey, 1987). These hypotheses illustrate the wide range of proposed origins for the HIV, from outer space to a Petri dish, from Columbus to bestiality. Admittedly, these are some of the more exotic hypotheses that have been posited for the origin of HIV, this "strange virus of unknown origin."

In this chapter we briefly review geographic dimensions of three of the more prominent hypotheses and related evidence pertaining to the search for the origins of the HIV; identify difficulties in each theory and the potential contribution of geographic research in this area; and present a general research strategy.

THE VIRUS AND THE SEARCH

In the scientific community—but for some skeptics, whose dissent is acknowledged here (Duesberg, 1987, 1989)—there is general acceptance of the human immunodeficiency virus as the disease agent underlying AIDS. Since the early stages of the epidemic, and particularly since the recognition of the pandemic of HIV infection, one of the central issues has been the search for the origin of the virus.

Only by conducting a search for its origin and accumulating evidence can the history and geography of the infection be more completely elaborated and demonstrated. More importantly, the search is vital to understanding the evolution and transmission of the HIV and how to control the biological and social mechanisms of the virus (McClure & Schulz, 1989; Essex, 1989). The prevention of AIDS is the real challenge and understanding the geographic origins of the HIV and the mechanisms and geographic dimensions of its transfer may eventually contribute to its control. The question persists: Whence did it come?

To date, the origin of the HIV remains a scientific mystery. As with other diseases such as venereal syphilis, the search for the origins of the HIV will continue for years, and we may never know its actual origin with complete assurance. Nevertheless, it is important to make the effort. The search is not without its difficulties and complications. Throughout much of history, ethnocentricity, enflaming political and social passions, and animosity have underlain assigning blame for the origins of various diseases, especially those carrying the social stigma associated with "aberrant" or "deviant" sexual behavior.

Certainly in the past a certain level of ethnocentricity was evident in attempts to locate and in statements describing the "index location" of a particular disease. In the 18th and 19th centuries, for example, yellow fever was known as the malady of Siam. During the syphilis pandemic in Europe in the late 15th and early 16th centuries, the Turks called it the disease of the Christians, to the English it was the French pox, to the French the Neapolitan disease, to the Italians the Spanish disease, and to the Spanish it was known as the Evil of the Island of Hispaniola brought to Spain by Columbus's crew upon their return from the New World in 1493 (Winslow, 1980; Andre, 1987). As we shall see, in some early as well as a few recent statements pertaining to the origin of the HIV there is evidence that the situation is not much changed. Certain groups of people and nations are blamed for somehow "originating" the virus. Frequently, in response, those being blamed will marshall "evidence" to either deflect the accusation or lay the blame at the doorstep of others. Inevitably, the process thereby becomes contentious.

Further compounding the problem of searching for the origin of the HIV, especially in the early stages of the pandemic and continuing in certain areas, is the lack of cooperation by certain nations and the denial of the existence of HIV-related health problems in some countries. In addition, some early testing procedures to detect the presence of antibodies to the HIV in stored sera—the major basis for designating and defining an index region—were later found to be imprecise, providing a rather high percentage of false indications of the presence of HIV antibodies. This has necessitated the reassessment of earlier statements regarding the distribution of the virus. Finally, the virus itself as well as associated conditions and diseases continues to evolve, leading to periodic redefinition of the virus indicator conditions. The clinical expression of infection appears increasingly complex. The types of opportunistic infections and neoplasms may vary not only in populations of different geographic origin but also according to the way the HIV infection was acquired (Piot & Colebunders, 1987). Therefore the distribution of HIV infection indicated by the presence or absence of certain presumed related conditions and diseases changes also.

Despite these obstacles, it is important to assess and attempt to integrate contemporary information pertaining to the geographic origin and diffusion of the HIV, acknowledging of course that, as our knowledge increases, conclusions regarding the geographic origins and pathways may change.

GLOBAL PATTERNS OF INFECTION

Though fewer than ten years have elapsed since identification of the pandemic of HIV infection on a global scale, three broad but distinct patterns of infection have been distinguished (Von Reyn & Mann, 1987; Mann, 1988; Torrey, Way, & Rowe, 1988).

Pattern I

The first pattern is found largely in sub-Saharan Africa and increasingly in Latin America, especially the Caribbean. HIV infection is thought to have begun to spread extensively during the mid-1970s. However, sexual transmission is predominantly heterosexual and the male-to- female ratios of infection are approximately 1:1. In this pattern the spread by intravenous drug use is relatively rare, but the virus may be spread by the repeated use of needles without sterilization and the common use of other skin-piercing instruments for medical or ritual purposes.

Because of the increased percentage of women infected, perinatal transmission is a major problem in some areas.

Pattern II

The second pattern is presently found throughout North America, Western Europe, Australia, New Zealand, and many urban areas in Latin America. It is characterized by the likely origin and extensive spread of the HIV during the late 1970s. Most cases occur among homosexual and bisexual males and intravenous drug users (Torrey et al., 1988). Heterosexual transmission, representing a small percentage of cases, is increasing. Blood products for transfusion are screened and essentially safe. Perinatal transmission, from mother to child, is uncommon because of the relatively few women thus far infected.

Pattern III

The third pattern is presently distributed in Eastern Europe, the Middle East, North Africa, and most countries of Asia and Oceania. Here, HIV infection appears to have been introduced relatively recently, that is, during the early to mid-1980s. Currently these countries account for only a small percentage of AIDS cases, less than 1% of the reported total. Early infections as well as large percentages of people developing AIDS are generally associated with transfusions of blood products from other areas, especially Pattern I countries. An additional source of infection appears to derive from sexual contact with populations from Pattern I and II countries—prostitutes are among the highest risk groups.

It should be noted that these patterns are broad generalizations and different patterns may coexist within a single country, or even within a single large metropolitan area (Figure 3.1). Additionally, the patterns can be expected to change as the infection spreads through the populations of these countries (Mann, 1988).

Though certainly not excluding any single country or region from consideration as an index location, these global patterns

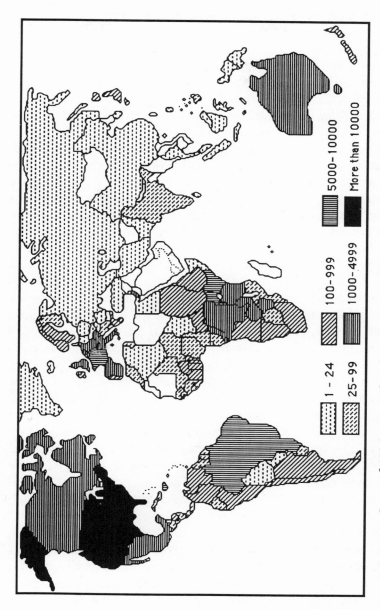

FIGURE 3.1. Cases of AIDS reported to the World Health Organization by the beginning of 1990.

suggest that our focus be directed away from Pattern III countries and concentrate our efforts on countries reflecting Patterns I and II. And indeed attention has been directed toward selected countries and regions within each pattern, namely: Haiti, Euro-America, and equatorial Africa.

THE HAITIAN CONTEXT

In 1981, practically coincident with the report of the first cases of AIDS in the male homosexual community in the United States, there were reports of 34 cases of AIDS among Haitian immigrants to the United States and 12 cases of a disease previously unrecognized in Haiti, an aggressive form of Kaposi's sarcoma (Barry, Mellors, & Bia, 1984). Moreover, AIDS was identified among both Haitian men and women, and the men were reportedly not homosexual. Quite naturally, in addition to the initial attention directed toward the United States, Haiti was identified as a potential index location for the HIV. And several rather disparate cases have been made for a Haitian origin of the HIV.

As mentioned earlier, one historical scenario traces the origin of the HIV to Haiti through the travels of Christopher Columbus (Andre, 1987). In addition to news of the discovery of a new world, it is conjectured that the crew of Columbus also brought back to Europe from Hispaniola (Haiti) a very contagious and "terrifying" disease manifest by cutaneous lesions, a deterioration of the general state, and "amputation of the genitals." Traditionally, this information has served as a basis for the "Columbian" or "American" theory of the introduction of venereal syphilis into Europe and the subsequent pandemic of the late 15th and early 16th centuries.

A second and considerably more modern Haitian scenario involves the spread of the HIV or its predecessor to human hosts in Haiti, as well as in equatorial Africa, through the regular ingestion of uncooked animal blood sacrificed in spirit-possession ceremonies (Moore & LeBaron, 1986). In this case, the focus is directed toward Haiti rather than equatorial Africa,

since the subsequent spread of the epidemic to Western populations, especially to the male homosexual communities in the
United States, was believed to occur thorough international
homosexual male tourism to Haiti, not central Africa (Figure
3.2).

Though nominally and ostensibly Roman Catholic, it has
been suggested that the large majority of Haitians participate in
the voodoo cult, a religion with a pantheon of gods derived in
large part from Dahomey in West Africa. Spirit possession of
priests is an essential element of the voodoo religion. In Haiti,
the spirits of the gods are summoned by the blood sacrifice of
bulls, goats, pigs, pigeons, and most commonly chickens. if accurately described, the priests, in some type of trance and
possessed by a patron saint, cut the throats of the animals or, in
the case of chickens, heads are torn off. The priest may ingest
the blood directly or the blood sacrifice is collected in a "gourd
containing salt, ash, molasses or native rum" (Metraux, 1958;
Moore & LeBaron, 1986). This mixture is stirred, allowed to
coagulate, and then the priest and his assistants ingest a portion.
Also, in healing ceremonies, the blood from the sacrificed
animal may be rubbed on or into a patient's afflicted part. Certainly this practice provides a frequent opportunity for exposure
to animal blood which may contain a precursor to the HIV
virus, or possibly the HIV virus itself.

This religious practice is in turn connected with a reported
high incidence of homosexuality or bisexuality among the male
voodoo priests (Metraux, 1958). In this regard historical accounts of Haitian culture depict the attitude toward homosexuality as "one of derision rather than vindictiveness" and male
homosexuality is present in both the cities and countryside
(Herskovits, 1971).

The next sequence in this theory involves homosexual contact
between Haitians and tourists from North America and the high
level of sexual activity among the latter during a period of the so-
called gay liberation movement (Leibowitch, 1985; Moore &
LeBaron, 1986). More specifically, in the 1970s, Port-au-Prince
developed as a popular resort area for homosexual men from
the United States. Given the documented high incidence of mul-

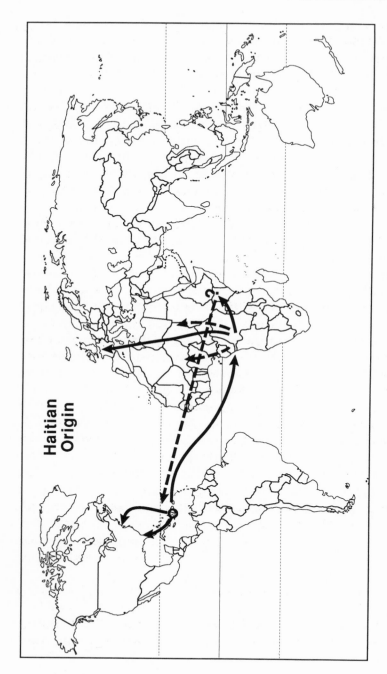

FIGURE 3.2 Possible AIDS diffusion pathways with Haitian origins.

tiple sexual partners among some homosexual men, it is hypothesized that the infection could have spread very rapidly to the tourists and, in turn, to the tourists' homosexual contacts upon their return to the United States.

It is proposed that the infection also traveled from Haiti to central Africa, most notably to Zaire (formerly the Belgian Congo) via the migration of a substantial number of middle-class Haitians in the mid-1960s. Their travel was a result of the recent achievement of independence by Zaire, granted by Belgium without adequate preparation. The official language of the Belgian Congo had been French, and to staff the civil service and administrative posts left vacant by the departure of the Belgians, Zaire recruited several thousand French-speaking Haitians. It is assumed the Haitians brought the HIV infection with them.

From Zaire the infection is presumed to have spread across central Africa and to Europe—to the latter by black Africans as well as white Europeans with strong ties to central Africa, and by Haitians who migrated to Belgium and France.

In summary, the Haitian scenario involves: (1) the presence of HIV-related viruses in animals (possibly spread from African animals to those of the Caribbean); (2) the use of these animals in blood sacrifices associated with the voodoo religion; (3) direct ingestion of this blood by *houngans* or voodoo priests; (4) the practice of homosexuality by priests and the tolerance of homosexuality and its practice among the general male population; (5) development of Port-au-Prince as a resort for male homosexuals from the United States; (6) high levels of sexual activity among some male homosexuals; (7) the migration of Haitians to central Africa; and (8) the migration of black Africans and Haitians from central Africa to Belgium and France.

ARGUMENTS AGAINST A HAITIAN ORIGIN

Haiti is not without its defenders. Some suggest that the disproportionate number of Haitians diagnosed as having AIDS may be due to the intense surveillance of a "captive" population such as the Haitian refugees in the United States. It is also suggested that the outbreak of AIDS in Haiti may be magnified by the opportunistic infections and generally widespread immunosuppression among Haitians resulting from chronic protein-calorie deprivation endemic on the island (Barry et al., 1984). Some also cite the difficulty in documenting risk factors such as sexual promiscuity, homosexuality, and drug abuse, the latter two illegal (Leonidas & Hyppolite, 1983; Smith, 1983). Also in defense of Haiti, some suggest the virus was introduced into the Caribbean from the United States by the male homosexual and bisexual community (Aubry, 1989).

Still other evidence has been presented against a Haitian origin for the HIV, such as: (1) the recency of the appearance of Kaposi's sarcoma in Haiti; (2) previous homosexual and heterosexual contact between Haitians and foreigners without complication; and (3) lack of involvement of hemophiliacs in the United States prior to 1975. The latter argument is especially interesting.

Hemophiliacs require repeated transfusions of coagulant Factor VIII derived from plasma. Apparently, from about 1970, concentrates of this factor were used in the United States, derived from the mixed plasma of from 2,000 to 20,000 donors. It is now documented that, prior to 1980, there is no recorded trace of HIV infection or related conditions from any of the case records or autopsies of hemophiliacs in the United States. This is especially important to the Haitian hypothesis, since prior to 1975 much of the blood used in North America to develop the concentrates came from Latin America and the Caribbean, namely, Haiti (Leibowitch, 1985). After 1975, the Federal Food and Drug Administration would no longer grant approval to blood from Haiti or the majority of former South American suppliers. Therefore, it can be concluded that prior to 1976 Haitian donors/sellers were not carriers of the HIV or a related HTLV.

Other information points to a recent development of the infection. A review of records of cancer biopsies from three private hospitals in Port-au-Prince, with a combined total of 180 beds, revealed no recorded cases of KS during the period from

1968 to 1983. And a review of 1,000 cancer biopsies from rural hospitals in Deschapelles during the same period also reveals no cases of KS. The earliest known patient with possible AIDS in Haiti appears to be a young man who died in 1978 (Pape & Johnson, 1989).

If in fact the virus was imported to Haiti, at least two sources of the infection are likely, namely, the United States and Africa.

A EURO-AMERICAN ORIGIN?

Despite the early identification of HIV-related conditions in the United States, the large numbers of reported AIDS cases in the United States and Western Europe, and the rejection of the "doctors of death" and bestiality hypotheses, there are currently few serious advocates of a Euro-American origin of the HIV (Figure 3.3).

In the very early phases of the epidemic speculation centered on the United States and its male homosexual population. Indeed, early on, the homosexual stamp was affixed to the disease, at first called gay-related immune deficiency (GRID). Attempts were made to link AIDS and AIDS-related conditions (ARC) with mutant immunosuppressor sperm that was the result of homosexual hypersexual activity (Leibowitch, 1985). Of course, it was not long before identification of AIDS and ARC in hemophiliacs, parenteral drug users, and other nonhomosexual individuals led to the discarding of this theory. And identification of ARC and AIDS in Haiti, central Africa, and Europe broadened geographic speculation on the origin of the disease agent.

Rejecting bestiality, germ warfare, and some type of recombinant biological research accident, it is difficult to develop plausible mechanisms for the origin of the HIV in humans or the transmission of an HIV precursor to humans from some animal reservoir in Europe or the United States. Perhaps prematurely, some immediately dismiss the United States and Europe from consideration by suggesting that AIDS conditions could not have broken out in these regions on an epidemic scale

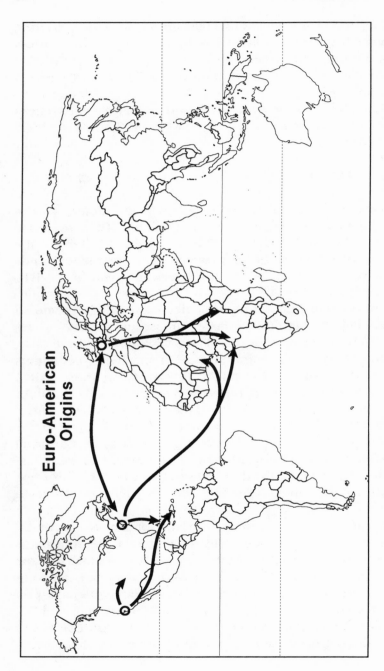

Euro-American Origins

FIGURE 3.3 The spatial diffusion of AIDS from American and European epicenters.

before 1980 without having been noticed. Therefore, barring some major genetic mutation, it is improbable that the HIV was somehow created "on the spot."

Similar to the proposed Haitian origin, however, some who posit a Euro-American origin of the HIV look to the morphologic and morphogenetic similarities between this virus and the visna-maedi virus found in sheep and goats (Najera, Herrera, & Andres, 1987; Kantner & Pankey, 1987). It has been proposed that the HIV is not a new virus and did not come from Africa, but has been endemic in the Euro-American population since the beginning of the 20th century. In one instance evidence to support this theory includes some 28 cases of disseminated Kaposi's sarcoma reported in the medical literature between 1902 and 1966.

How the virus spread to humans from animal reservoirs poses an even greater problem to the authors of this hypothesis. They make but do not follow up on a statement that sexual contact between male humans and sheep has been documented, presumably implying that this was the mode of transmission. Moreover, they discuss the formation of sado-masochistic clubs and an increase in traumatic sexual practices, again without explaining the linkages to the spread of the disease but suggesting that sociocultural changes associated with the gay liberation movement caused the infection to become epidemic and somehow exported to Africa. From another source it is suggested that the HIV was exported to Africa in blood products imported from Europe and the United States (Sabatier, 1987).

One Euro-American origin scenario proceeds as follows: (1) a virus closely related to the HIV was present in the sheep of northern Europe; (2) the virus was somehow transmitted to humans through human male sexual contact with the sheep; (3) the viruses lay dormant for many years until (4) increased promiscuity and traumatic sexual practices developed among homosexuals in the 1970s; and (5) the HIV was transmitted to Africa in blood and blood products imported from Europe and the United States.

To date, there exist few comprehensive studies and, therefore, little evidence in support of this Euro-American hypothesis.

Nevertheless, in some countries of Africa and Asia, the generally accepted view is that the HIV spread from Europe to Africa rather than the other way around, or, coupled with anti-Western sentiment, it began with American male homosexuals and was spread around the world by American travelers and imported American blood products (Waite, 1988; Sabatier, 1987).

OUT OF AFRICA?

Attempts at retrospective identification of early HIV infection and AIDS cases in some countries of Africa as well as countries in other developing regions is problematic for a number of reasons (Hayes, Marlink, & Hardawi, 1989). For example: (1) Because resources are extremely limited, clinical record-keeping has a low priority. (2) Where there is clinical evidence suggestive of HIV infection or AIDS, the necessary corroborative laboratory evidence may be lacking. (3) The lack of resources may not permit diagnosis of AIDS by means of invasive techniques used in developed countries. (4) Immunologic testing of fresh and stored sera is difficult. There are reports of false positive confirmations by Western blot tests from fresh sera and it is realized that cross-reacting retroviruses may be circulating in tropical and other regions (Quinn & Mann, 1989). Some older sera show immunologic reactivity despite the absence of the HIV. It is probable that the enzyme-linked immunosorbent assay (ELISA) testing yields a significant number of false positives, especially in areas of endemic malaria. Ideally there would be additional specific tests to confirm all repeatedly positive ELISA tests.

Amid these problems and caveats and continued published claims of a conspiracy among scientists and the media against Africa (Versi, 1990), as well as persistent denials of the existence of a major health problem, there are several lines of evidence that suggest the HIV-1 and the more recently identified HIV-2 may have originated in sub-Saharan Africa (Konotey-Ahulu, 1987; Farthing, Brown, & Staughton, 1988; Chiodi, Biberfeld, Parks, et al., 1989; Hayes et al., 1989; Essex, 1989) (see Figure

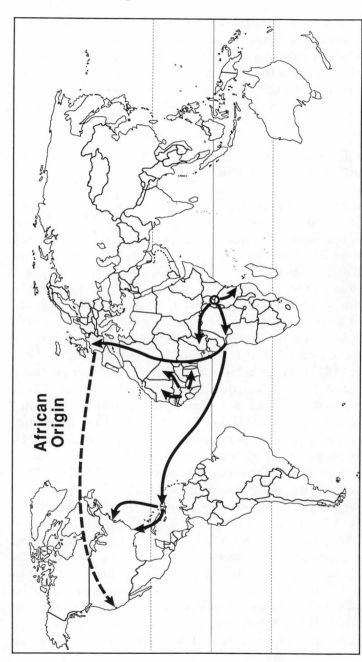

FIGURE 3.4 Postulated international diffusion pathways of AIDS during early phases with African origins.

3.4). The eastern and central regions of Africa—Burundi, Kenya, Rwanda, Tanzania, Uganda, Zaire, and Zambia in particular— are particularly important in this regard because of their proximity to the epicenter of the African AIDS outbreak and because of their relatively high reported incidence of AIDS (Yeager, 1988).

SERO-EPIDEMIOLOGICAL AND
SERO-ARCHEOLOGICAL EVIDENCE

Using more refined and reliable tests and Western blot criteria (JAMA, 1989), corroborated sero-epidemiological and sero-archeological studies indicate the presence of HIV-1 infection in Africa as early as 1959 in a serum sample collected from Leopoldville, Belgian Congo, now Kinshasa, Zaire (Nahmias, Weiss, et al., 1986; Chiodi et al., 1989). In 1970 a pregnant woman was found to be HIV-1 seropositive in Kinshasa (Brun-Vézinet, Rouzioux, Montagnier, et al., 1984). Clinical records also indicate a suspected case of AIDS in a Danish surgeon apparently exposed to HIV-1 in Zaire between 1972 and 1975 (Bygbjerg, 1983). Examination of sera collected in a northern Zairian equatorial region during investigations of an Ebola River fever outbreak indicates the presence of the HIV-1. Additionally, an isolate of the HIV-1 was recovered from the serum of one antibody-positive person who subsequently died of an AIDS-like illness in 1978 (DeCock et al., 1987). Finally, a recent study of sera collected from several remote tribes in Zimbabwe, Liberia, and Kenya in the late 1960s and early 1970s confirmed the presence of the HIV-1 in two specimens from the Mano tribe of Liberia (Chiodi et al., 1989).

With the notable exception of KS and other mysterious symptoms found in the frozen tissue of a 16-year-old black American male who died in 1969, no serum samples stored in the United States prior to the 1970s have been found to be seropositive. On the basis of serological studies there is some indication that HIV infection may have emerged earlier in Africa than in the United States. However, the rapidly rising incidence of cases in Africa also suggests a new epidemic of infection, perhaps as recent as 40 to 50 years ago (Farthing et al., 1988; Essex, 1989; McClure & Schulz, 1989).

SIMIAN IMMUNODEFICIENCY VIRUSES

One hypothesis suggests a primate reservoir of a precursor of HIV, the African green monkey. Wild African green monkeys in captivity have a very high incidence (approximately 60%) of infection with an immunodeficiency virus closely related to HIV called simian T-lymphotropic virus type III or simian immunodeficiency virus (SIV) (Farthing et al., 1988). This virus does not appear to cause AIDS in the African green monkey, but when injected into Asian macaque monkeys from a different continent it results in AIDS-like diseases and conditions.

However, it should be noted that the SIV is much more closely related to a second, more recently discovered strain of the HIV, namely HIV-2, than it is to HIV-1 (Essex, 1989). The HIV-2 has been located in several countries of western Africa including Cape Verde, Burkina Faso, Ivory Coast, Guinea-Bissau, and Senegal. (Kanki, Alroy, & Essex, 1985, 1987; Clavel, Mansinho, & Chamanet, 1987; Horsburgh & Homberg, 1988; Essex, 1989). It is reported that the class of reactivity seen with serum samples from West African prostitutes was virtually indistinguishable from that seen with serum samples from sooty Mangabey monkeys.

Since the HIV-2 and SIVsm are highly related to each other, and the HIV-2 is apparently restricted primarily to West African people, whereas SIVagm is present in green monkeys throughout much of Africa, the possibility that this virus moved from monkeys to people can obviously be considered. The SIVs, HIV-2, and HIV-1 are clearly related, and each is more closely related to the other than to other lentiviruses and the human retroviruses HTLV-I and HTLV- II.

In any event it seems that SIVagm in African green monkeys was not the immediate precursor of HIV-1. Therefore, one hypothesis might be that HIV-1 moved into humans from a precursor virus in another as yet unidentified species or, alternatively, that there has been substantial genetic drift in the HIV-1 within the human primate. A more direct relationship appears to exist between the HIV-2 and the SIV derived from captive and wild sooty mangabey monkeys in captivity (SIVsm) indigenous to West Africa (Hirsch, Olmsted, Murphey-Corb, Purcell, & Johnson, 1989).

A sample of HIV-2 obtained from a serological survey of Senegalese prostitutes appeared to be almost serologically indistinguishable from the SIVsm and appears to be more logically considered an immediate precursor of the HIV-2 than the SIVagm is for the HIV-1. Moreover, the HIV-2 appears to be endemic among several West African countries. In so far as there may be two distinct types of the HIV and their geographic location is also distinct, it follows that there may be two distinct geographic foci of the SIV and HIV development within Africa. Alternatively, one might suggest a horizontal development of the SIVs among nonhuman primates with subsequent transmission of HIV precursors to humans.

With regard to the etiological origin of HIV infection among humans, therefore, the most cogent issues derive from the initial isolation of the SIV from the African green monkey and subsequently from the sooty mangabey. Several tentative hypotheses have been posited to explain the possible route of transmission from monkeys to humans, including insect vectors such as mosquitoes, monkey bites, or the consumption of monkey meat.

Given the historical epidemiological experience of diseases such as yellow fever, malaria, and the bubonic plague, perhaps the most intuitively plausible thesis is the transmission of the pathogen by biting insects (Zuckerman, 1986). Although it has been demonstrated that the HIV-1 can survive for from several hours to several days in insects fed or injected with blood containing high concentrations of HIV-1 (Lyons, Schoub, & McGillivray, 1985), epidemiological evidence from Africa as well as the United States weighs heavily against the role of a blood sucking arthropod as a vector. HIV-1 infection is rare in children, otherwise known to be risk-free, who are bitten most frequently by mosquitoes, fleas, lice, and the like. Also, no significant correlation between the presence of HIV-1 antibodies in humans and arboviruses in high incidence areas has been established (Biggar, Melbye, et al., 1985; Mann, 1987b; Castro, Leib, et al., 1988). Moreover, the concentration of the HIV-1 in the blood of infected humans is quite low, further reducing the likelihood of transfer by biting insects.

Regarding the consumption of monkey meat, there is no substantive evidence to support the spread of the HIV-1 via the enteric route. Finally, the relatively low concentration of the HIV-1 in saliva renders the bite of an infected monkey quite an inefficient route of HIV transmission.

Certainly the possible origin of HIV in humans, whether HIV-1 or HIV-2, remains in dispute (Essex & Kanki, 1988; Mulder, 1988), and no scientifically proven model exists for the method of transfer. Nevertheless, there are certain practices extant in Africa that could bring humans in direct contact with blood from monkeys. Interestingly, one such practice has been documented among natives living in remote central Africa near the presumed epicenter of the African epidemic. In documenting the sexual customs and culture of people living in the Great Lakes region of central Africa in 1973, Kashamura (1973) describes the following practice of the Idjwi, who reside on an island in Lake Kivu, between the borders of Rwanda and Zaire:

" . . . pour stimuler un homme ou une femme et provoquer chez eux une activité sexuelle intense, on leur inocule dans les cuisses, la région du pubis et le dos du sang prélevé sur un singe, pour un homme, sur une guenon, pour une femme." (To stimulate a man or a woman and induce in them intense sexual activity, monkey blood [for a man] or female-monkey blood [for a woman] was directly inoculated in the pubic area and also in the thighs and back.)

If accurately described, such a practice would constitute a very efficient means of transferring the SIV to both males and females. Through genetic mutation, such a transfer could be in part responsible for the emergence of the HIV-1 and HIV-2 in humans. It might also serve to corroborate one model which suggests that HIV infection may have existed and remained stable in remote areas of central Africa for a long period of time (Nzilambi, DeCock, et al., 1988).

Further support for the long-standing presence of HIV infection among natives in remote areas derives from recent evidence of HIV-1 infection among the Sangha pygmy group in the isolated ecosystem of the Central African Republic (Gonzalez, Georges-Courbot, et al., 1987) and the first cases of AIDS in Uganda occurring among several businessmen who died at an isolated fishing village on Lake Victoria (Okware, 1987).

Though complex and not completely defined, currently, the African origin scenario couples the diffusion of the infection on the continent with the corresponding period of rapid urbanization and international trade: specifically, migrant labor systems, the migration of young men to cities, the development of the infection in prostitutes, and prostitutes' mobility as well. Of particular importance in this scenario is the current highway transportation system and the Trans-Africa Highway which is proposed to link Mombasa, Kenya on the east coast to Lagos, Nigeria, on the west coast. Though far from completed, this road passes through Kenya, Uganda, and Rwanda (near Lake Kivu), to northern Zaire, the Central African Republic, Cameroon, and Nigeria. Reports from Africa suggest that HIV-1 infection appears to be spreading eastward from central Africa along the Trans-Africa Highway (Okware, 1987).

If the initial reservoir of HIV infection is central Africa, several geographical pathways out of Africa have been suggested. Some countries of Western Europe have long established colonial ties with Africa and substantial concentrations of central African immigrants. For example, an estimated 6000 to 8000 residents from central Africa, mostly from Zaire, reside in Belgium (Clumeck, Sonnet, et al., 1984). Many frequently return to Zaire for family or other business purposes while others come to Europe for medical care. In fact, with the description of AIDS cases from the United States in 1981, clinicians particularly in Belgium and France recognized a similarity between these cases and patients arriving for medical evaluation from the central region of Africa (Biggar, 1987). Also, some early cases of ARC and AIDS in France occurred among immigrants recently arrived from central Africa (Brunet, Boubet, & Leibowitch, 1983; Ellrodt, Le Bras, et al., 1984).

In this scenario, rather than moving from Haiti to Zaire, it is posited that the Haitians recruited to fill civil service posts in the immediate post independence period acquired the infection in Zaire and carried the infection with them as they traveled back to Haiti or to Europe (DePerre, Lepage, et al., 1984). Subsequently, primarily through international travel of homosexuals, the infection traveled from both Haiti and possibly Europe to the United States. In turn, the United States and Europe served as reservoirs for further diffusion of the virus to other countries of the world.

In summary, the African origin theory involves: (1) the presence of HIV precursors in the primate population throughout most of central Africa; (2) transmission of this precursor from an animal to human host; (3) diffusion throughout central Africa supported by rapid urbanization, migration, and transcontinental transport; (4) diffusion to Haiti by Haitians returning from employment in Zaire; (5) diffusion to Europe via migrating Africans and Haitians from central Africa; (6) infection of male homosexuals from the United States visiting Haiti; and (7) infection of populations in other countries by visiting Americans or the importation of infected blood products.

DIFFICULTIES WITH
THE AFRICAN ORIGIN THEORY

Several issue have been raised regarding the African origin theory. While some can be dismissed rather readily, others warrant close consideration. For example, if the AIDS epidemic due to the HIV-1 were the outcome of transmission from monkeys, it is puzzling that it has not been possible to document infections with the HIV-2 or other SIV-like viruses among populations in central Africa (Kitchen, 1987). The widespread distribution of African green monkeys throughout equatorial African countries, the close relationships between the HIV-2 and SIV, and the absence of the HIV-2 from eastern Africa, the purported epicenter of HIV-1 infection, are particularly puzzling.

Further, the HIV-2 has been identified as cytopathic and capable of causing AIDS. The HIV-2 is closely related to the SIVsm found in the sooty mangabey monkeys indigenous to West Africa. Is it possible there are two (or more) types of HIV precursors located in different species of monkeys that were somehow almost simultaneously transmitted to humans in two geographically distinct areas of sub-Saharan Africa? The situation may be much more complex and speculative than currently described.

However, recent evidence indicates that extensive genetic variation of the SIV exists in African green monkeys from a single geographic region (Naidu, Daniel, & Desrosiers, 1989). This extensive diversity justifies a continued search for isolates more closely related to the HIV-1 and HIV-2 that would provide evidence for cross-species transmission between African green monkeys, sooty mongabey as well as other monkeys, and humans as the origin of HIV syndrome viruses.

Another argument against the African origin of HIV suggests that, if Africa is the source of the infection why was the syndrome first identified in American homosexuals, and not in Africa? The answer to this question may be related to the lack of appropriate diagnostic facilities in Africa to detect an emerging syndrome as polymorphous as HIV-related disease, and to the early recognition of cases in the United States due to their concentration in limited groups at risk rather than diffusion over the general population, as is not the case in Africa (Desmyter, Surmont, Goubau, & Vandepitte, 1986).

CLOSURE?

Which origin theory, if any, is correct? The search continues. To date, there exists no conclusive scientific evidence for locating the exact origin of the HIV. As we have seen, the search for the origins of the HIV has exhibited an unusual propensity to exacerbate existing tensions within the body politic (Sabatier,

1987; Versi, 1990). Amid claims and counterclaims, for example, efforts of Western researchers to locate the origins of the HIV in Africa have been denounced as a conspiracy and perpetuation of racially motivated stereotypes. Tensions exist even within Africa. It is reported that some Nigerian athletes refused to attend the All Africa Games in Nairobi because of the prevalence of AIDS there (Schmidt, 1988). Certainly, caution must be exercised when interpreting data on HIV infection from any location.

The search for the possible origins of the HIV does indicate however that, regardless of geographic region, we are woefully lacking in our understanding of behavior patterns important to the transmission of the HIV. Our knowledge of sexual practices appears to be as lacking and fragmented for the United States as it is for Zaire and Haiti. For example, we have no accurate data on geographic variations in either intra- or international homosexual practices. Legal sanctions against homosexuality in some countries and social sanctions in others certainly affect the accuracy of informants' response to any such inquiries.

In a similar fashion, we do not have accurate information on the sexual practices of heterosexuals, regardless of geographic location. If, as reported, anal intercourse is a particularly efficient means of transmitting the virus, it would seem essential to understand the incidence of this sexual practice among heterosexuals as well as homosexuals. Transmission of the virus via oral sex among male homosexuals was originally suggested several years ago (Evans, McClean, Dawson, et al, 1989), and a recent case appears to support the efficacy of this pathway (Spitzer & Weiner, 1989). In this instance a man is believed to have been infected by a woman.

HIV infection is spread through different types of behavior, and presently the best hope for stopping the epidemic is through changing the types of behavior responsible for its continued transmission. Yet human behavior and the forces that shape this behavior are among the most complex and poorly understood dimensions of the problem. The knowledge base in the behavioral and social sciences necessary for a search for HIV origins is rudimentary at best.

Therefore it is incumbent upon social scientists to establish the necessary knowledge base. It is also obvious that the various social sciences and medical sciences cannot work toward this goal in "splendid" isolation. The problem is too complex and we do not have the luxury of unlimited amounts of time. It is time for increased cooperative and consolidated efforts within the social sciences, among anthropologists, geographers, psychologists, sociologists, and others; cooperation between social and medical scientists; and cooperation between both groups and representatives of persons living with HIV infection.

Perhaps reflecting our bias as geographers, we would suggest further that the search for the origins of the HIV as well as efforts to develop strategies to modify related behavior patterns proceed on a spatial and regional basis. The regional tradition is well established and continues strong within area studies, anthropology, and geography. We should build on these traditions and develop regionally specific interdisciplinary research teams in cooperation with regional governments adequate to conduct the necessary sociobehavioral research.

4

The African Experience

Apart from the interest in Africa as a possible index location for the origin of the human immunodeficiency virus, the focus of world attention has been directed toward other aspects of the African experience with the HIV and AIDS. Estimates from several sources suggest that the number and proportion of people infected with the HIV in many countries of central sub-Saharan Africa are as great or greater than in any other region of the world. These estimates of HIV-1 infection range from 2 to 5 million (Miller & Rockwell, 1988). Also, the potential destructive impact of AIDS on the population structure and economic development of many countries in Africa is considered to be among the most threatening to date (Prewitt, 1988). As mentioned in Chapter 3, there are some who question whether these observations represent the best and most objective considerations of the situation or are the extravagant journalistic hyperbole and untruths resulting from a "conspiracy [among the] world's media and some scientists" (Konotey-Ahulu, 1987).

It is not certain whether the African experience would draw the concern and attention of so many were it not for the almost unique distribution of HIV infection throughout the population.

In central Africa the infection appears almost equally distributed between men and women. The approximate 1:1 male-
to-female ratio of HIV infection observed in this region contrasts sharply with the ratios of from 20:1 to 15:1 observed thus
far in most developed countries of the world affected by the HIV
epidemic. In the absence of any substantive evidence to the con-
trary, the HIV appears to spread bidirectionally predominantly
through heterosexual intercourse (Piot & Mann, 1987). Perhaps
motivated as much by self-interest and self-preservation rather
than purely humanitarian reasons, the majority of Western
research seems to focus on attempting to explain and understand this different and potentially very threatening pattern of
HIV infection found in central Africa and the surrounding
regions. In any event, today the problem of AIDS does appear to
be more complex and potentially more serious in Africa than on
any other continent of the world.

Many of the countries in Africa with the greatest potential for
destruction from AIDS suffer from other problems as well. For
example, political violence in Uganda has severely damaged a
once thriving rural economy. In Zaire bureaucratic inefficiency
and corruption arise from and feed on ethnic and sectional instability. Rwanda is faced with rural overcrowding and natural
resource degradation so extensive they have been characterized
as "terminal." In Zambia, the collapse of copper revenues and
near bankruptcy prevent economic recovery and investment in
a long neglected agricultural sector. Finally, in Kenya an estimated 80% of the world's fastest growing population is com-
pressed into less than 20% of the available land (Yeager,
1988).

Superimposed on this framework, AIDS creates a problem of
enormous medical, social, and geographic complexity. As long
as there is no known cure for the disease, the only way to avoid
the risk of infection is to avoid exposure. In Africa and increasingly elsewhere, especially because HIV transmission appears to be largely heterosexually transmitted, reducing exposure means changing a basic biologically driven behavior. In
many instances culturally as well as socially specific attitudes
toward sexual relationships and related behaviors conducive to

HIV infection are condoned and even sanctioned by cultural traditions.

According to Zambian tradition, for example, when a husband dies his male relatives must have sex with his widow to "purge" his ghost. Among senior women, there is a strong resistance to any move to eradicate this practice. Though eradication would reduce the risk of HIV transmission, it would also, they feel, reduce the opportunity for widows to remarry (Sabatier, 1987). Among the Lese of Zaire, during the period following puberty and before marriage sexual relations between young men and a number of eligible women are virtually sanctioned by the society. In the so-called matrilineal belt centered in south central Africa, there is an especially high degree of adolescent promiscuity and uncertainty about paternity. Many patrilineal African societies are promiscuous as well (Hrdy, 1987).

In this sense, again, AIDS is a "behavioral" illness. Changing or modifying sexual and other behaviors on a broad scale may be impossible due to the plethora of cultural and societal value systems. This situation precludes any single, homogeneous educational program. These complexities are underscored for Africa when we realize there can be no panacea for a continent that is four times the size of the United States, with over 900 ethnic groups and 300 language families distributed over 58 countries (Miller & Rockwell, 1988).

To date there is no definitive statistically significant study of the number of HIV-infected persons in Africa, or elsewhere for that matter. Efforts to determine the number of current and past cases are frustrated on several fronts. First, it must be realized that in many if not most central African countries, AIDS is still considered far down on the list of health problems, superceded by diarrheal disease, malaria, measles, tuberculosis, and cholera, to name a few (Dickson, 1987a; Mann, 1987). To illustrate, in Uganda, though the problem of AIDS is severe, it is relatively minor when compared to other health problems. This country has a population of about 17 million, approximately 8 million of whom are 14 years of age and under. In 1987 it was estimated that 600,000 children die of preventable diseases compared to 300 reported deaths among the 2,369 reported cases of AIDS (Gaq, 1988). If estimates are correct and only 1 in 50 AIDS cases in Uganda is currently being seen and/or reported by health authorities (Good, 1988), the health impact of AIDS is growing but may still be considered relatively minor compared to other health threats. The situation may be changing, for by the end of October 1989 Uganda had the highest reported number of AIDS cases in Africa—7,735. This figure could translate into a potential 400,000 Ugandans infected with HIV. In all likelihood, Zaire may have the highest number of AIDS cases in Africa. But, due to continuing political and economic instability and corruption, accurate figures are not available (Yeager, 1988). In the latest report available (WHO, 1990) a total of 4,636 cases of AIDS had been reported to the World Health Organization from Zaire by the end of 1989. The current figure, though unreported, is certainly much higher.

Attempts to identify AIDS cases in Africa are made difficult for several reasons:

1. Keeping clinical records has a low priority as health resources and personnel are inadequate.
2. In instances where clinical records exist and are suggestive of AIDS, corroborative evidence may be lacking.
3. Techniques necessary to diagnose AIDS-related diseases may not be available.
4. Available tests for HIV antibodies may yield a significant number of false positives (Hayes, Marlink, & Harawi, 1989)

With regard to the identification of HIV infection and AIDS itself, early serological studies were confused by the presence of an unexpected high frequency of nonspecific reactivity and false-positive indications of HIV infection, but in studies done since 1986 this problem has been resolved by better technology (Kuhls, Nishanian, Cherry, et al., 1988). In addition, a revised protocol for the identification of AIDS was developed for Africa at the Bangui (Central African Republic) Conference in 1985 (Airhihenbuwa, 1989; WHO, 1986). This case definition of AIDS, based on the Bangui Workshop Criteria, is more attuned to the clinical realities and lack of resources so prevalent throughout much of Africa. The case definitions for adults and children are based on a common enteropathic version of AIDS known as "slim disease" because of cachexia or emaciation associated with diarrhea. The definition includes both major and minor criteria, as illustrated in Table 4.1.

In addition to basic problems in identifying HIV infection and related opportunistic diseases and conditions, relatively little information is available on the behaviors related to the diffu-

TABLE 4.1. WHO Clinical Case Definition of AIDS
(Bangui Workshop Criteria)

AIDS is defined by the existence of at least two of the major signs associated with at least one minor sign, in the absence of immunosuppressive diseases such as cancer or severe malnutrition or other recognized etiologies.

Major signs	Adults	Children
Weight loss > 10% of body weight	X	X
Chronic diarrhea > 1 month	X	X
Prolonged fever > 1 month		
(intermittent or constant)	X	X

Minor signs		
Persistent cough for >1 month	X	X
Generalized pruritic dermatitis	X	X
Recurrent herpes zoster	X	
Oropharyngeal condidiasis	X	X
Chronic progressive and/or		
disseminated herpes simplex	X	
Generalized lymphadenopathy	X	X
Repeated common infections		
(otitis, pharyngitis)		X
Confirmed maternal HIV infection		X

The presence of disseminated Kaposi's sarcoma or cryptoccal meningitis are sufficient by themselves for the diagnosis of AIDS.

Source: Hayes et al. (1989), Airhihenbuwa (1989), and Von Reyn & Mann (1987).

sion of the HIV in Africa. There has been little concerted effort
to study the full range of sexuality and health in sub-Saharan Africa generally, or central and East Africa specifically. Information about sexual beliefs, behaviors, and related illness conditions is at best widely scattered (Barton, 1988). Only relatively
few studies describe sexual practices and sexually related diseases among the wide variety of ethnic groups. While such
a tradition may be defended as culturally sensitive and pre-
vents stigmatizing particular groups, it renders the information
base incomplete and the extent and usefulness of available information limited.

Despite these problems, a considerable body of information
has been developed and assembled on the African experience
with HIV infection and AIDS. Though the information is admittedly incomplete and changing, it is useful to piece together this
information in an attempt to characterize the African experience.

INTERNATIONAL GEOGRAPHIC PATTERNS

By the end of 1989, with the exception of Libya, Madagascar,
Mauritania, and Seychelles, all countries in Africa reported at
least one case of AIDS (WHO, 1989b). None of the four countries has reported an AIDS case as of February 1990 (WHO,
1990). To emphasize what was only alluded to above, all countries in Africa are not equally affected by HIV infection and
AIDS, and it is incorrect to speak of or consider the "African experience" as implying some type of homogeneity across the
continent (see Figure 4.1). Some nations are relatively free of
AIDS and others seem to show case levels of infection and
associated conditions several times higher than in developed
nations (Miller & Rockwell, 1988).

The eastern and central regions of sub-Saharan Africa—
Burundi, Kenya, Rwanda, Tanzania, Uganda, Zaire, and Zambia
in particular—are notable because of their proximity to the apparent epicenter or, better, epicenters of the African AIDS
epidemic and their relatively high, though still perhaps vastly

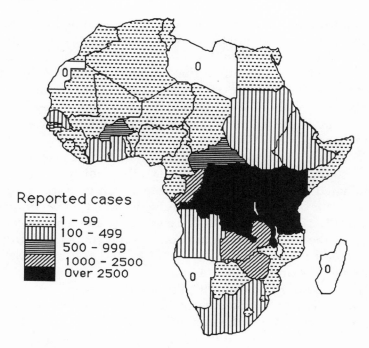

FIGURE 4.1 Cumulative numbers of cases of AIDS reported to the World Health Organization from African countries to January, 1990.

underreported incidence of the disease (Yeager, 1988). Their aggregate estimated 1989 population was 117 million, of which 74% was rural and 26% urban. The countries themselves are by no means uniform, ranging in size and population from Burundi, about the size of Maryland and with a population of just over 5 million, to Zaire, about one-fourth the size of the United States, with a population of 34 million.

Prior to 1989, Zaire reported 335 cases of AIDS. In Kinshasa alone, however, 332 cases were identified from a serological survey conducted between July 1984 and February 1985. Moreover, a "reasonable" estimate of the true incidence was between 550 to 1,000 cases per million adults. At that time, Kinshasa had a population of approximately 3 million people (Quinn, Mann, Curran, & Piot, 1986). It should be emphasized here that the underreporting of AIDS cases is certainly not limited to African countries. This is believed to be a common practice throughout much of the Middle East, Eastern Europe, and the U.S.S.R. It may also be more common than supposed even in countries such as the United States. In South Carolina, for example, it is estimated that about 40% of the AIDS cases went unreported in a one-year period prior to mid-1987 (Conway, Colley-Niemeyer, Pursley, et al., 1989).

There is also extreme ethnic variation between and within these countries most affected by AIDS. Burundi, for example, is dominated by the Hutu ethnic group which comprises about 85% of the population. In Zaire, while the Bantu comprise about 80% of the population, there are some 200 separate tribes residing there, but individual countries' percentages of urban population range from Rwanda (5.1%) and Burundi (8%) to Zaire (44.2%) and Zambia (49%).

At the end of February 1990, the seven countries comprising the AIDS "belt" accounted for 30,512 reported cases of AIDS— approximately 74% of the total of 41,516 cases reported for all countries of Africa. In addition to these "core" countries, high levels of confirmed cases of AIDS are reported from countries in an extension of the "belt" westward through the Congo, Ghana, and the Ivory Coast, as well as south through Zimbabwe and Malawi (Good, 1988). When these five countries are included, the "epidemic belt" countries of central and eastern sub-Saharan Africa account for 38,067 reported AIDS cases, or 92% of the total number of cases reported for all of Africa.

Five of these countries—Burundi, Malawi, Rwanda, Tanzania, and Uganda—are among the 42 poorest nations in the world. According to statistics compiled by the United Nations conference on Trade and Development, the average annual per capita income is about $200. In fact 28 of these poorest countries are located in Africa including the sub-Saharan countries of the Central African Republic, Equatorial Guinea, Gambia, Guinea, Guinea-Bisseau, Mozambique, Mali, Mauritania, Niger, and Togo (Crossett, 1990).

By February 1990, 48 countries had reported more than one case of AIDS. Based on the distribution of AIDS cases represented by these data, it is very apparent that the countries of central Africa and adjacent areas are currently most severely affected by the HIV, while populations in most of the north, west and southern Africa have a relatively lower incidence of HIV sero-positivity and fewer number of identified AIDS cases.

Though the patterns are striking, one still must be cautious in drawing any "hard and fast" conclusions based on data present-ly available. The reporting is very uneven from country to coun-try and the actual identification of AIDS cases is problematic in many countries and actively suppressed in others (Barker & Turshen, 1986; Kitchen, 1987). Further, HIV of one type or another is now being detected in areas previously thought to be free of infection. In some of these areas it is difficult to dis-tinguish whether the HIV was recently introduced or the virus was only recently recognized.

Important here is the apparent differential distribution (see Figure 4.2) of a relatively recently discovered second type of HIV, labeled HIV-2. As described earlier, the HIV-2 is much more closely related to the simian immunodeficiency virus than is the HIV-1. As previously noted, the class of reactivity seen with serum samples from Senegalese prostitutes was virtually indistinguishable from that seen with serum samples from sooty mangabey monkeys or infected captive rhesus monkeys (Essex, 1989).

The screening of more than 2000 high-risk individuals from central Africa, including many with AIDS and other sexually transmitted diseases (STDs), revealed no evidence that a virus more related to the SIV was present in that area. Conversely, screening in West Africa revealed that 1% to 10% of the control adults in countries such as Senegal, Guinea-Bissau, Gambia, Burkina Faso, and Ivory Coast had evidence of HIV-2 infection. (Essex, 1989; Mabey, Tedder et al., 1988) To date, the HIV-2 ap-pears to be endemic only in West Africa and, based on analysis of stored sera, it may have been present there since 1966 or earlier (Horsburgh & Homberg, 1988; Kawamura, Yamazaki, Ishikawa, et al., 1989; Odehouori, DeCock, Krebs, et al., 1989).

Recent evidence points to the endemicity of the HIV-2 in West Africa, and an exceedingly high degree of similarity between the

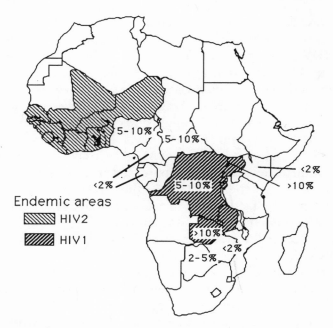

FIGURE 4.2 Area considered endemic for HIV-1 and HIV-2 in Africa along with estimates for HIV-1 seropositive proportions of populations.

HIV-2 and the virus of the sooty mangabey monkey (SIVsm) indigenous to West Africa (Hirsch, Olmsted, Murphey-Corb, Purcell, & Johnson, 1989). Though admittedly the sequence of infection remains to be demonstrated, these observations together with the apparent pathogenicity of HIV-2 related to AIDS suggests there may be, in fact, two geographical epicenters of HIV infection in Africa. As mentioned previously, the major center of the HIV-1 appears to be the Great Lakes Region of central-eastern Africa. Now, in addition, a second epicenter linked to the HIV-2 may be found in West Africa.

At the same time these possibilities are considered, however, the African origin theory is further complicated by the almost simultaneous transmission to humans of the HIV-1 precursor from the African green monkey, or some other primate, in central and eastern Africa and the HIV-2 precursor from the sooty mangabey monkey in West Africa as discussed in a previous chapter.

NATIONAL PATTERNS

There is tremendous variation in the international experience with AIDS in Africa, both between and within affected countries. In the recent stages of the epidemic the major distinction in most countries has been the predominance of AIDS cases identified in urban rather than rural areas. However, this is not necessarily the case for all African countries. In Rwanda and Zaire cities do appear to be the most affected (Mann, 1987). For instance, in Rwanda 94% of the population is rural, but only 4% of AIDS cases diagnosed in a 1983 surveillance study were from rural areas. Moreover, the seroprevalence rate in Rwanda's capital of Kigali (17.5%) is almost six time higher than that found in a sample from one rural area (3.0%) (Torrey, Wai, & Rowe, 1988).

A study conducted in the remote Equateur province of Zaire found less than 1% of human serum samples collected in 1976 tested positive for HIV. In 1986, a serosurvey in the same area found essentially the same seroprevalence rate (Nzilambi, DeCock, Forthal, et al., 1988). On the other hand, in the capital city of Kinshasa HIV incidence rates for adults are comparable to those found among Haitian immigrants to the United States and one-third the rate among intravenous drug users in New York City (Mann, Francis, Quinn, et al., 1986). In 16 years the seroprevalence rate among pregnant women in Kinshasa increased from less than 1% to 8% (Johnson & Laga, 1990).

On the other hand, in Uganda, the first AIDS cases were suspected during the latter part of 1982 among several businessmen who died at Kansensero, an isolated fishing village on Lake Victoria. The small town is a reputed center for smuggling and other illicit business transactions. Among the local populace, the initial opinion was that the deaths were simply due to witchcraft or "natural justice" against cheaters. Soon, however, there were corresponding illnesses and some deaths among the spouses of the men. The social stigmatization attached to the illness forced some of the families to move to larger inland towns which were believed to be previously unaffected by the disease. From this remote village setting, AIDS has now become an urban disease in Uganda.

The initial theorized path of the infection was from the isolated fishing village to nearby urban areas, and from those areas it spread eastward to other towns and cities along the major transportation routes, the Trans-African highway in particular. The situation in 1987 is presented in Figure 4.3. It is estimated that 33% of the long-distance truck drivers in Uganda were infected at the time (Okware, 1987). In Uganda the number of reported AIDS cases has increased from 17 in 1983 to over 7,500 in early 1990.

The first reported AIDS cases in Tanzania occurred in the rural northwest region of the country, and enteropathic AIDS was first identified near the Tanzanian border in rural Uganda. By late 1989 over 4,000 cases of AIDS had been reported to the Tanzanian Ministry of Health and Social Welfare, and still more than 60% were from the northwest region of the country. Seroprevalence remains highest in this region. Among pregnant women in the capital city of Dar-es-Salaam the estimated rate was 3.6% in 1986, and in northern Arush, pregnant women had a seroprevalence rate of 0.7%. However, a survey of pregnant women in Bukoba, a large Tanzanian town near the Ugandan border, estimated a seroprevalence rate of 16% (Torrey et al., 1988).

In Kinshasa, Zaire early seroprevalence studies of women attending an antenatal clinic in Kinshasa indicate that less than 1% of the population tested positive in 1970, suggesting that the virus was then only rarely present in places that would be classified as cities within the AIDS belt of central and East Africa (Biggar, 1987; Essex, 1989). By 1980, 3% were seropositive and by 1986, 7% were positive.

Again, these observations lead to the speculation that either the virus moved from nonhuman primates to people just prior to the identification of AIDS or that it had been transmitted to

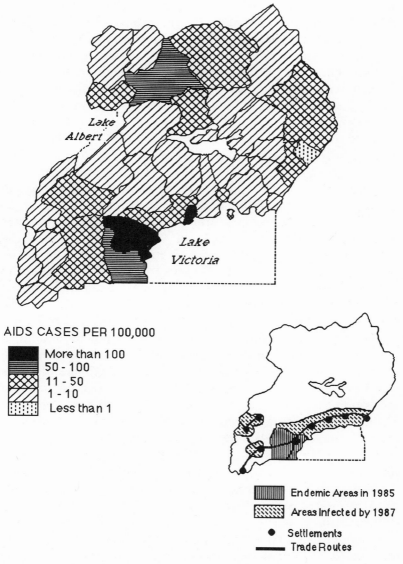

AIDS CASES PER 100,000

- ■ More than 100
- 50 - 100
- 11 - 50
- 1 - 10
- Less than 1

- ‖‖‖ Endemic Areas in 1985
- Areas Infected by 1987
- ● Settlements
- — Trade Routes

FIGURE 4.3 The cumulative incidence of AIDS in Uganda and the spread of HIV-1 along trade routes during the 1980s (adapted from Okware, 1987).

cities by the migration of a few resistant carriers from a previously isolated tribe or tribes. The latter might be supported by
evidence that rural areas constitute a large and important reservoir of STDs as well. In many parts of eastern central Africa
rural populations are sufficiently affected by the prevalence of
STDs to reduce fertility rates to below rates elsewhere in Africa
(Barton, 1988).

GEOGRAPHIC ORIGIN AND DIFFUSION

Although the discovery and apparent endemicity of the HIV-2 in
West Africa confuses the issue, as mentioned earlier, attention is directed toward central Africa as the index location
of the HIV-1 and, hence, AIDS in Africa. Here too, however,
lack of data precludes definitive statements.

With the description of AIDS cases from the United States in
1981, clinicians in Europe, particularly Belgium and France,
recognized similarities between these cases and patients arriving for medical evaluation and treatment from the central
region of Africa (Biggar, 1987, 1988). Patients with similar illnesses were soon found in Zaire and Rwanda, the native countries of many of the African patients with AIDS-like illnesses
seen in Europe. However, there was some initial skepticism that
the conditions were similar because the profile of the patients,
1.2 males to 1.0 females, was vastly different from that found in
the early stages of the epidemic in the United States, where the
male-to-female ratio was 19:1. Subsequently, the discovery of
the HIV-1 as the causative agent demonstrated that the syndromes were in fact similar and caused by the same virus in Africa, the United States, and Europe.

The distribution of AIDS cases in Africa points generally to a
region of east central sub-Saharan Africa as one possible index
location for the origin of the pandemic on the continent. More
specifically, from several sources, the so-called Great Lakes
Region is a focal point of speculation (see Figure 4.4). This
region includes the largest lakes in Africa, Lake Victoria, Lake
Tanganyika, and Lake Nyasa, as well as smaller ones such as

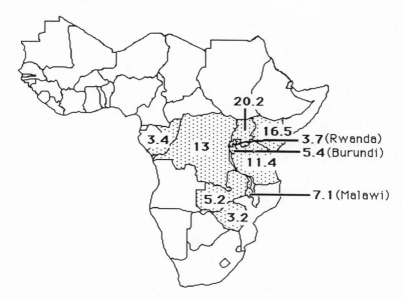

FIGURE 4.4 The shaded areas are countries considered part of the African "AIDS Belt." Percentages are based on the total numbers of African cumulative cases reported to the World Health Organization by January, 1990.

Lake Kioga, Lake Rudolf, Lake Albert, Lake Edward, and Lake Kivu. The list of countries that contain or border on these lakes includes the countries in which the largest numbers of AIDS patients have been documented to date. Bordering Lake Victoria, for example, are Uganda, Kenya, and Tanzania. The countries of Zaire, Burundi, Tanzania, and Zambia surround lake Tanganyika. Malawi, Tanzania, and a small section of Mozambique border lake Tanganyika. The smaller lake system including Lakes Albert, Kivu, and Edward comprise part of the borders of Uganda, Zaire, Kenya, Rwanda, and Burundi.

The high concentration of AIDS cases in this region leads to one theory that the spread of the immunodeficiency virus in Africa is contigual and possibly emanated from a single or several central locations. A seroprevalence study in rural Zaire compared serum samples from 1976 with those from 1986 (Nzilambi, DeCock, et al., 1988). The prevalence in the remote area was low (0.8%) in 1986 but believed to be stable since 1976.

One major implication is that the HIV may have originated among populations of the more remote areas of central Africa and was transferred to the urban areas with migrants moving to the cities. It is possible that the virus or its precursor was confined to remote areas and only recently spread to major cities in Africa (Montagnier & Alizon, 1987). Biologically it would make sense if the infection had started in a small rural area, especially if, as indicated, the reservoir for the HIV or its precursor occurs among nonhuman primates of the forests and jungles (Desmyter, Surmont, Goubau, & Vandepitte, 1986).

Generally, as might be expected, the prevalence of AIDS cases and HIV seroprevalence is significantly lower in rural populations (Hayes, et al., 1989). In part, this apparent urban bias may be due to data derived from the more modern and better equipped urban hospitals and the greater ease with which seroprevalence studies may be conducted among selected urban populations. On the other hand, even among urban hospitals there is evidence of substantial use by residents of surrounding rural areas. In an earlier study of gonorrhoea in Kampala, Uganda, for example, half the cases treated in hospital came from surrounding rural areas and the radius of daily movement came from the city extended to a radius of some 30 miles (Bennett, 1962).

The current status of information pertaining to the origin of the HIV as well as data reporting the number and distribution of AIDS cases preclude definitive conclusions on the origin and diffusion of the virus and syndrome. Nevertheless, the accumulated literature to date strongly suggests the original source of the infection among humans was located in some of the more remote areas of central East Africa, especially in the Great Lakes area. From here, infected individuals may have migrated into larger towns situated along major highway and waterway transportation routes. Infection rates appear to be rising rapidly in market towns situated on major trade and labor migration routes (Schoepf, Nkera, & Schoepf, 1988). Subsequently, the infection developed rapidly in previously unexposed populations in these towns and cities, so that now in most countries urban populations are most severely affected.

Of particular interest here is the role played by prostitutes living in the cities, or migrating from city to city, and one segment of their male customers, comprised of recent rural–urban or seasonal migrants and transport workers involved in international trade (Clumeck, Van De Perre, Carael et al., 1985; Torrey et al., 1988).

Prostitutes and their customers in all countries heavily affected by AIDS appear to have the highest HIV rates of any groups studied. In Ghana prostitutes are estimated to have five times the rate of HIV infection of others (Torrey et al., 1988). Among urban prostitutes in Uganda, the HIV infection rate is estimated to be over 80%. Surveys conducted in 1987 indicated a 13% HIV seropositivity among pregnant women in Kampala, 33% among long-distance truck drivers, and 85% among prostitutes in the Rakai District (Haq, 1988).

In Nairobi, the HIV-1 was detected in the serum of 66% of "low class" prostitutes and 31% of "high class" (Kreiss, Koech, Plummer, et al., 1986). No significant association was found between presence of the HIV-1 antibody and the women's age, length of prostitution experience, number of customers per year, nationality, history of immunization, parenteral medications, transfusions, scarification, or several other factors. However, there was a significant association between presence of the HIV-1 antibody and sexual exposure to men of several different nationalities. The relative risk (RR) of the presence of the HIV-1 antibody after sexual exposure to partners from Rwanda was 8.73, Burundi 5.0, and Uganda 4.76. It is interesting to note that while many of the prostitutes were from Tanzania and Uganda, there was no association found between presence of the HIV-1 antibody and nationality (Kreiss et al., 1986).

Evidence from prostitutes in the Republic of Djibouti located on the horn of Africa indicates that the HIV is becoming a serious problem there (Fox et al., 1989). In only a nine-month period, from October 1987 to June 1988, the HIV seroprevalence rate among prostitutes increased from 1.25 to 2.6%.

POPULATION MOVEMENT

Population movement is one of the most significant themes in the study of the HIV and AIDS in Africa, for it may be through this means that the HIV has moved into the larger cities and has been carried from one region to another. In part, population migration creates social conditions conducive to behavior that causes the transfer of the virus. Who moves or migrates? Why? When?, and For how long? are important questions to be explored if the diffusion of the infection is to be understood and any significant attempt at control is to be made (Brokensha, 1988). As with other aspects of the African experience, population movements are of many types and the picture is complex. A summary of the major dimensions is presented here.

Some of the conditions facilitating the spread of HIV infection can be traced to the development of the colonial economic structure in many African countries. A comparison of the political map of Africa today with that of a relatively short time ago reflects both the depth of colonialism on the continent and the only recent and continuing emergence from this period (see Figure 4.5).

In the late 19th century, there was intense competition among Belgium, Portugal, Great Britain, Germany, and France for the possession of central and eastern African territories. By 1920, tens of thousands of African males and, in lesser numbers, females traveled to rural European estates to work as farm laborers, to mining regions to work as miners, or to new colonial cities to work as domestics or in shops and factories, or to sell agricultural produce (Dawson, 1988).

While some of the movements were permanent, many men moved without their families to take up lengthy, albeit temporary, residence while working in the cities. From a very early date, therefore, the colonial African urban population had a very skewed male-to-female sex ratio. For example, as early as 1911, Nairobi had a sex ratio of six males for each female. Long absen-

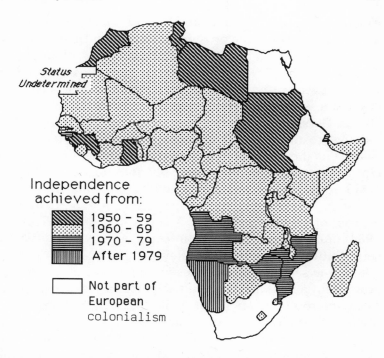

FIGURE 4.5 Time periods independence gained from European colonialism powers of the nineteenth and twentieth centuries.

ces from home almost inevitably led men to seek female companionship with prostitutes, who began to practice their profession at a very early age in most major colonial cities.

While political independence has brought some stability to the labor force in most central and eastern African countries, many urban workers retain strong ties to rural areas, and there is a great deal of movement to and from cities. Today, the population structure of some of the larger cities still resembles that of colonial cities as the rapid urbanization begun in the 1970s continues.

In so far as they play a major role in the spread of HIV infection, today prostitutes are also a significant migrant labor force. As mentioned earlier, for example, many prostitutes in Nairobi,

Kenya come from Tanzania and Uganda. In coastal Mombasa, an estimated 20% of the prostitutes were from Uganda and 40% of all prostitutes had lived in Mombasa for less than one year (Dawson, 1988). In Tortiya, northern Ivory Coast, 68% of the prostitutes surveyed for seroprevalence had practiced in this city for less than two years and 27% for less than one year (Denis, Gershey-Damet, Lhuillier, et al., 1987). Further, almost half the prostitutes were from neighboring Ghana. It is also suggested that the frequent movement of prostitutes between larger cities and smaller towns is probably not unusual.

Moveover, the infection rate among prostitutes is also increasing. In 1981 sera were collected from 116 prostitutes in Nairobi of which 5 (4%) were seropositive. Another survey in 1984 found 174 of 286 (61%) infected, and by 1986 85% were seropositive (Piot, Plummer, Reyn et al., 1987; Anonymous, 1989). The seroprevalence rates among prostitutes varies geographically across Africa as well. Reported rates include Nigeria 1%, Cameroon 7%, Mali 15%, Ivory Coast 20%, Zaire 38–46%, Malawi 50–62%, and Tanzania 54–77%. Comparatively, the epidemic among prostitutes elsewhere is only beginning, with estimated seropositive rates among prostitutes in the Philippines, .1%; Thailand, .2%; Mexico, .2%; India, 2.6%; Brazil, 6.2%; and Puerto Rico, 2.7%. However, as discussed earlier the infection rate may increase rapidly as evidenced by the experience in Thailand where HIV infection rates from between 13 and 70% have been found among prostitutes in some major cities. There is evidence that the situation in some Caribbean countries is as severe as that in Africa. In Martinique some 43% of the prostitutes are infected as well as 4% of Haitian prostitutes there.

Another mobile population traditionally implicated in the transmission of the STDs in Africa, and HIV infection more recently, is truck drivers. During the colonial period, they were accused of helping to spread venereal syphilis to western Kenya. In East Africa today there is concern over the role they play in the spread of HIV infection as they travel between large urban areas such as Nairobi, Mombasa, Kampala (Uganda), and Kilgali (Rwanda). Of course, the drivers also stop in smaller urban centers to make deliveries or rest, frequenting local bars, hotels, prostitutes, and other "free" women. These men also have families and ties to rural areas (Dawson, 1988).

If, as suggested, long-distance truck drivers serve as a vector for transmission of the HIV, then of particular concern must be potential route offered by the Trans-African Highway. As originally conceived, the Trans-African Highway will link Lagos, Nigeria, on the Atlantic coast of Africa with Mombasa, Kenya, on the Indian Ocean to the east. Passing through a considerable portion of Africa already heavily affected by the virus, the highway and its "tributaries," once completed, would comprise a comprehensive road system opening up previously isolated areas to intraregional development and providing land-locked countries access to ocean ports and international trade. As proposed, the highway would cross Kenya, Uganda, northern Zaire, the eastern Central African Republic, northern Cameroon, and southern Nigeria (see Figure 4.6).

Throughout history population movements have been associated with the diffusion of infectious diseases, and AIDS is certainly no exception. In Africa, population movements are large, frequent, and continuing, contributing to the sexual integration of various groups. The entire central and eastern African area continues to experience large shifts of population. There is long-term movement of rural populations into urban areas and frequent return migration. Other, more recent trends include the movement of migrant workers from Zaire and Rwanda to neighboring countries, movement of armies on the Uganda–Tanzania border, and larger numbers of refugees, especially from Uganda (Hrdy, 1987). These population movements, in addition to those associated with national and international trade, probably have been significantly associated with the current patterns of HIV infection and AIDS. Of particular importance, however, is the potential of these population movements to support the diffusion of the HIV to major cities and large towns across international boundaries as well as into previously unaffected populations living in smaller towns and villages in rural areas.

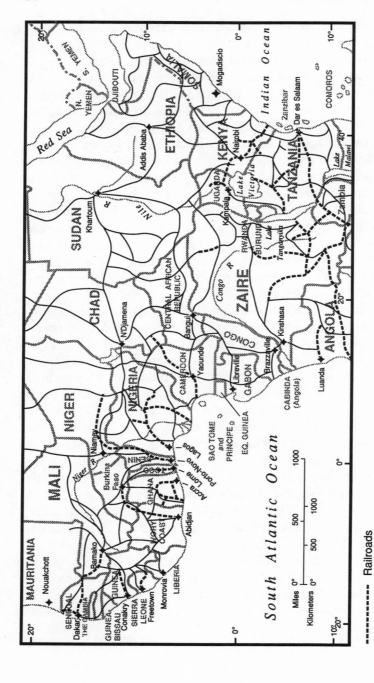

FIGURE 4.6 Major trans-African transportation routes.

ETIOLOGICAL ORIGIN IN AFRICA

Observations of HIV seroprevalence existing in remote rural populations in several countries of central Africa, and patterns of early AIDS cases originating among populations in remote villages, point to these areas as possible reservoirs of the HIV or its precursor. There is some additional support for the notion of a rural origin of the HIV and the presence of the HIV for some time among almost totally isolated population groups. In one such study, sera collected from three African tribes residing in Zimbabwe, Liberia, and Kenya were investigated for the presence of antibodies to both the HIV-1 and HIV-2 (Chiodi, Biberfeld, Parks, et al., 1989). The sera were originally collected from 1969 through 1971 in connection with a study of viral antigens. At the time of sera collection, the isolated living patterns of the tribes were relatively unchanged. Though conclusions are tentative, interestingly, antibodies to both the HIV-1 and HIV-2 were not found in samples from the Korekore and Turkana tribes from northern Zimbabwe and northern Kenya, respectively. However, evidence of HIV-1 antibodies was detected among specimens from the Mano tribe of northern Liberia, from an 85-year-old woman and 60-year-old man (Chiodi et al., 1989).

Also, HIV-1 antibodies were recently isolated in a serological survey of Aka and Babinga pygmy populations in remote forest sections of Lobaye and Sangha in the Central African Republic (Gonzalez, Georges-Courbot, Martin, et al., 1987). The antibodies were found in a 33-year-old healthy woman and her 40-year-old male sexual partner.

If, as indicated, the virus originated in rural or remote areas of central Africa, the question remains, What was the original source and how was it transmitted? Here, too, the situation is complex and gradually emerging. There are several species of "old world" African and Asian monkeys infected with SIVs, which are closely related to the HIV (Zuckerman, 1986; Letvin, Daniel, King, et al., 1988). There is some speculation that the HIV mutated from one of the SIVs (STLV-III), and currently most attention is directed toward the presence of a similar retrovirus in the African green monkey found distributed throughout central, eastern, and much of southern Africa (Li, Naidu, Daniel, & Desrosiers, 1989; Biggar, 1987) (see Figure 4.7).

From 30% to 70% of wild African green monkeys are infected with an SIV closely related to the HIV. Recent studies indicate that there is extensive genetic variability in the SIVagm, much greater than that observed in the HIV-1 and HIV-2. This suggests a relatively recent transmission to the human species, certainly less than 100 years and perhaps as recently as 40 years ago (McClure & Schulz, 1989; Essex, 1989). Moreover, it is quite possible that an SIVagm or an SIV from some other species of monkey or ape even more closely related to the HIV-1 may be isolated, providing additional support for the theory of an

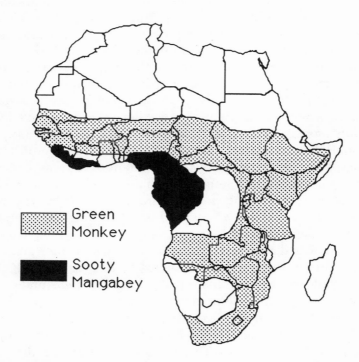

FIGURE 4.7 The distributions of Green and Sooty Mangabey monkeys in Africa.

evolutionary relation between the human and simian viruses (McClure & Schulz, 1989; Li et al., 1989).

In any event, the SIVagm does not appear to cause AIDS-like symptoms in the green monkey, but does when injected into Asian macaque monkeys (Farthing, Brown & Staughton, 1988). It should be mentioned here that related SIVs have been identified in monkeys of Asia, but those from Japanese macaques or related Asian species are believed to be significantly less related to the HIV and the SIVs isolated from African primates (Letvin et al., 1988; Essex, 1989). Further, all isolates of the HIV, whether from Japanese, Caribbean, or African people, are highly related to African strains of the SIV, but not as highly related to Asian strains. This suggests that all HIVs thus far identified probably evolved from a subgroup of the SIV present in Africa but not in Asia. Most important, it also suggests that the SIV/ HIV family of viruses has been present in numerous species of old-world monkeys for some time before moving to man from an African species of monkey, chimpanzee, or ape.

Since the SIV does not apparently cause AIDS-like disease in its monkey hosts, we can conclude that the SIV and its host have apparently undergone the host–parasite evolutionary selection process, which fosters the development of a relatively attenuated virus. Therefore, rather than just host immunity, if this virus was then transmitted to man, it would be a virus of low virulence. It is speculated that an SIV of low virulence might have been transmitted to man, where it may have remained a virus of limited virulence for some time. Or, if a virulent HIV-1 or HIV-2 had been present in population subgroups in Africa to the point of evolutionary equilibration as seen with the SIV, it probably would have been limited to isolated tribes of people, and would represent a situation in which selection for host immunity rather than selection for an avirulent virus would have occurred (Essex, 1989). We might speculate then that either the virus moved from subhuman primates to people relatively recently or that a few resistant carriers from a previously isolated tribe or tribes transmitted the virus to the cities by migration.

Cultural and Societal Cofactors

Are there other practices or conditions that may contribute to the widespread and almost equal HIV infection rates among men and women in a basically heterosexual society? Several have been suggested, including (1) promiscuity, with a high prevalence of sexually transmitted disease; (2) sexual practices presumed to be associated with increased risk of HIV transmission such as bisexuality and anal intercourse; and (3) cultural practices that are possibly connected with increased virus transmission such as male circumcision and female "circumcision" and infibulation (Hrdy, 1987).

Promiscuity

As early as 1985, it was suggested that the spread of the HIV infection among heterosexual populations in Africa might be related to the heterosexually promiscuous sexual lifestyle of some African men (Clumeck et al., 1985). A study of African men with AIDS in Kigali, Rwanda and Brussels, Belgium indicated that when compared to a healthy control population, the men had a significantly higher median number of different heterosexual partners per year and more frequent—at least once a month—contact with prostitutes. The number of sexual partners ranged from 12 to 60 per year and 81% had regular contact with prostitutes. Among the controls, 34% had regular contact with prostitutes and the average number of sexual partners per year was three.

A recent study also sheds initial light on the sexual lifestyle of African men in a pastoral African setting (Konings, Anderson, Morley, O'Riordan, & Meegan, 1989). Sexual partner change among unmarried male warriors in two tribal communities in East Africa over a period of three years ranged from 2 to 38 and averaged 12 per year. It is also reported that custom among these groups encourages the sharing of possessions including sexual partners. Compared to studies of male heterosexuals in the United Kingdom the number of partners among these African warriors was on the magnitude of 10 times as great.

A common thread of promiscuity should be noted here. In 1985, the Wellcome Institute of Medicine in London reported the average number of sexual partners of AIDS patients to be 1,160, about twice that of noninfected homosexuals. In this group, as in African male and female heterosexuals, promiscuity appears to be a cofactor in the risk of infection. The average number of different sexual partners per month for male homosexuals in selected countries has been reported to be: Finland 1.7, Ireland 2.0, Australia 2.7, and Sweden 3.3 (Ross, 1984).

Though no definitive information is available, it is suggested that male homosexuality among African men probably occurs on a minor scale compared to that found in the West. Nevertheless it does occur and cannot be completely dismissed as a factor in the transmission of the HIV (Brokensha, 1988). Reportedly, large percentages of African homosexual men live where there are also large numbers of males. In all-male compounds and prisons homosexuality is in greater existence. And, perhaps more important, historically in a number of African societies certain males assume female roles. The relevance of this practice for the transmission of the HIV may be important, depending on the number of sexual partners of these men.

Female Circumcision and Infibulation

Information available to date suggests that normal vaginal intercourse is a relatively inefficient process for HIV transmission. In this regard it has been proposed that heterosexual HIV-transmission is enhanced by female circumcision. An estimated 90 to 100 million women living in Africa have been subject to a traditional form of female mutilation. Several types of female circumcision are practiced in Africa (Fourcroy, 1983; McLean & Graham, 1985; Lightfoot-Klein, 1990). Infibulation, or pharaonic circumcision involves the partial closure of the vaginal orifice after excision of varying amounts of tissue from the vulva. In its extreme form, all of the mons veneris, labia majora and minora, and clitoris are removed and the involved areas closed by means of sutures or thorns. Complete occlusion of the introitus is prevented by the insertion of a matchstick or other wooden object. In a more moderate practice, excision involves the removal of the clitoris and part of the labia minora. The least traumatic form, sunna circumcision, involves removal of the clitoral prepuce. In many instances the vagina is almost completely sealed, leaving an opening large enough for urine to pass drop by drop. In order to have sexual intercourse, it is often necessary for repeated gradual penetration (over a period of 2 to 12 weeks), which is essentially a process of repeated tearing, to stretch the opening and in some instances the vagina must be cut open to facilitate sexual relations and childbirth. After childbirth, the vagina may be resewn, only to repeat the same process again.

It is hypothesized that these practices may increase the likelihood of female exposure to the HIV during intercourse by the tearing of tissues due to the small introitus, the presence of scar tissue, and the abnormal anatomy of a mutilated vagina. Alternatively, such a practice may lead to other forms of sexual intercourse, anal intercourse in particular, which would also subject the female to a greater risk of infection.

Geographically, however, the current focus and distribution of AIDS and HIV infection does not strongly correspond to the known distribution of the practice of infibulation (see Figure 4.8). The practice occurs primarily in the Arabian peninsula, isolated areas of West Africa, the horn of Africa, the Sudan, and northern Kenya. It has been stressed, however, that knowledge pertaining to this practice is very incomplete in large sections of the AIDS belt which encompasses much of eastern and central Africa.

Further, it is possible that large population movements from rural to urban areas and across national boundaries may have introduced female circumcision into urban areas where it was previously not practiced. For example, except for the Luo, the practice is still widespread in urban areas of Kenya. In any event, HIV infection and AIDS is now being reported in many countries where female circumcision is known to be widely practiced and the implications for exacerbating the spread of the epidemic are important. While the practice of female circumcision may enhance the spread of the HIV, the absence of

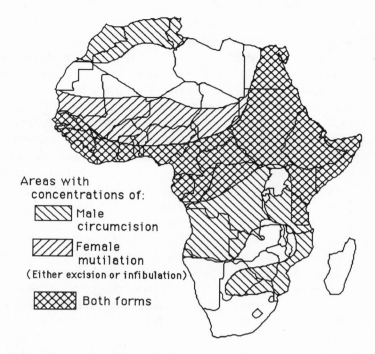

Areas with
  concentrations of:
  Male
    circumcision
  Female
    mutilation
(Either excision or infibulation)

  Both forms

FIGURE 4.8 Known areas where male circumcision and female mutilation are practiced.

male circumcision may be a crucial factor in the spread of the HIV in Africa.

Male Circumcision

Among some tribal and ethnic groups in Africa, indigenous cultural tradition requires men to be circumcised. Adherence to one of the major world religions is also important. Male circumcision is required of the followers of Islam but not for those who follow Christianity. The focal point of the geographic distribution of male circumcision in Africa occurs in the central sub-Saharan region (see Figure 4.8). In Rwanda and Burundi the percentage of males who are circumcised is estimated to be zero. Other countries have very low percentages, for example,

Zambia 2.5%, Malawi 4.6%, and Uganda 9%. The proportions of circumcised males rise in Tanzania (49%) and Zaire (75%) (Bongaarts, Reining, Way, & Conant, 1989). The geographic focus is again directed toward countries of the Great Lakes Region of central Africa. Geographic variations in the lack of circumcision occur even within countries. In western Kenya (around Kisumu located on Lake Victoria) northwestern Kenya, and western Tanzania, the practice is absent and HIV seroprevalence rates for these areas are the highest in each country. While exceptions to the geographic correlation between lack of circumcision among males and high rates of AIDS and HIV infection exist, especially in some urban areas, a strong association warrants further study, nonetheless.

Especially pertinent to this line of inquiry is the observed relationship between lack of male circumcision and genital and sexually transmitted diseases. It is generally recognized that venereal disease such as genital herpes and syphilis are more common in uncircumcised men (Taylor & Rodin, 1975; Parker, Stewart, Wren, Gollow, & Straton, 1983; Fink, 1986). Since the infectious agents of both these diseases depend on a break or abrasion in the skin to gain entry into the body, the cervical secretions of a woman infected with the HIV are more likely to be transferred by a similar means through the delicate, easily abraided penile lining, such as the mucosal inner layer of foreskin, than when the foreskin is absent. Among a Zimbabwean population, for example, wives of men with a history of genital ulcer disease were more likely to be seropositive. HIV infection was also associated with other sexually transmitted diseases. It was also observed that male-to-female transmission of the HIV-1 is facilitated by the presence of genital ulcers in infected men (Latif, Katzenstein, Bassett et al., 1989).

CONCLUSION

Africa is a complex and dynamic continent and subsequently so is the African experience with AIDS and HIV infections. Understanding the geography of Africa is essential to understanding the African experience. Perhaps most basic is understanding the "milieu" of Africa, its cultures and societies as well as its biogeography and medical geography. Against this background, the geography of historic and contemporary population movements must be traced at the international, national, regional, and local levels. We are not without knowledge of Africa in any of these subjects, but our information is relatively limited and from geographically scattered locales. The spatial and temporal perspective embodied in the geographic point of view and shared with other social science disciplines is essential to understanding more about the origin of the virus and its spread across Africa. More importantly, understanding the geography of the African experience can lead to the development of programs necessary to providing relief to those affected and restricting damage to many developing nations.

5

The Progression of
AIDS in Europe

Geographical aspects of HIV infection and AIDS in Europe are best understood in terms of three temporal phases and three kinds of spatial patterning. The combination of temporal phases and spatial patterning results in different directional paths and geographical clusters of countries or tracts. Based on the European experience to date, the temporal context suggested here includes: (1) an early seeding period that probably extended from the early 1970s to the early 1980s in countries with considerable African and U.S. contacts, and from the early through the late 1980s in many Eastern Bloc countries; (2) an infusion or consolidation stage wherein small numbers of confirmed cases were reported to and subsequently officially recognized by public-health authorities, resulting in increased frequency of reported cases; and (3) the "prime surge" phase, beginning in Western European countries during the late 1980s, exacerbated in part by a major nomenclature revision in 1987 which broadened the range of conditions and diseases presumptive of AIDS and resulted in a jump in reported cases.

The three types of international geographical patterning

began with the formation of a core area during the seeding and infusion phases of the infection. As the infection diffused and the number of reported AIDS cases as well as the number of countries affected increased in the 1980s, the differentially affected countries formed a core-fringe and periphery. At the time of this writing, the core-fringe consists largely of Western Europe, and the periphery extends to Eastern Europe and to such countries as Iceland, Finland, the Soviet Union, and Albania.

Clearly, several kinds of spatial diffusion processes have been operating almost simultaneously in Europe over the past decade or longer. The spread of HIV infection and AIDS has been hierarchical to the extent that large cities separated by considerable distances and connected by international airline routes operated as initial points for the infusion or entry of the HIV and probably the seeding of the virus as well. Although we may never know the complete details about the seeding process, there are indications that sexual promiscuity and tainted blood products may have been involved.

From these major entry points, intraregional and local diffusion followed, particularly during the prime surge phase. During this phase the number of countries representing the core area as well as the corresponding core-fringe areas expanded. Thus, the initial long distance point-to-point geographical diffusion of HIV infection and AIDS cases can be traced to routes of international airlines and related international travel. Several directional diffusion pathways or "tracts" of countries in Europe are suggested. The subsequent local and intraregional diffusion patterns can be identified with the classic core and cluster neighborhood-effect patterns common to other contagious and sexually transmitted diseases (Brown, 1981) on a larger geographical scale. Retrospective analyses of possible HIV-seeding in Europe help in understanding the possible origins of the disease in this part of the world.

EUROPEAN ORIGINS UNLIKELY

The initial isolation of the lymphadenopathy virus at the Pasteur Institute in Paris and the human T-cell lymphotropic virus type

III at the National Institutes of Health in Bethesda, Maryland, in 1983 determined the distant relationship between this human retrovirus and the others identified in the late 1970s (Weiss & Biggar, 1986). Also in 1983, a reexamination of the death in 1976 of a Danish physician working in Zaire revealed conditions consistent with the clinical definition of AIDS.

In Belgium, from 1979 through 1983, 17 Africans and one Greek who had lived in Zaire for 20 years were treated for AIDS-related opportunistic infections or Kaposi's sarcoma (KS). All were reportedly heterosexual, did not use intravenous drugs, and had received no blood transfusions during the five years prior to the onset of their health problems. In 1985 a British research team reported that HIV infection and AIDS were present in British male homosexuals and hemophiliacs prior to 1980 and, further, that the geographic origin of the infection was probably the United States (Mortimer et al., 1985). At the same time, an Italian report similarly concluded that, as early as 1980, "... HTLV-III may have entered Italy as a result of the spread in the United States" (Auti et al., 1985). There is, in fact, little doubt that the United States was a major source of HIV infection in Europe, especially during the infusion and prime surge phases.

At the same time, however, several countries of western Europe had substantial concentrations of migrants from central Africa, many of whom retained strong and frequent ties to this region. For example, in 1984 an estimated 6,000 to 8,000 people from central Africa, mostly from Zaire, lived in Belgium (Clumeck et al., 1984). Many returned frequently to Zaire for business or social/family reasons. In addition, many other Zairians traveled (and still do) to Belgium and other parts of Europe for medical care. It was among these people that a cluster of early AIDS cases was identified. Similarly, many early cases of AIDS and AIDS-related conditions in France occurred in people recently arriving from central Africa (Ellrodt et al., 1984; Brunet et al., 1983; Lancet, 1983). Another potentially infected pool consisted of whites who traveled to and returned from living in central Africa (Bygbjerg, 1983; Tauris & Black, 1987). By the spring of 1983, the 44 known cases of AIDS in West Germany occurred primarily among people who had visited either Haiti or Africa or among homosexual males who had visited the United States. In addition, the geographic patterns identified below provide additional evidence that the HIV did not originate in Europe but in both the United States and Africa.

THE EARLY PHASES OF AIDS IN EUROPE

As with many contagious diseases, certain areas are generally well-seeded with the disease agent before the outbreak or epidemic of the related disease is recognized. The delay is created by the amount of time required for confirmation of the cases and diagnostic corroboration. Frequently, once epidemic conditions are recognized, there is a period of intense and frequent futile retrospective activity in search of an index case. Most sustained efforts, however, are prospective in nature. There is often a time lag between recognition of the first few cases and official reporting. Thus, particularly during the early phases of an epidemic, tabulations of cases are generally sparse and lacing in geographic accuracy.

Generally, reporting improves with time and, despite still enormous underreporting, this appears to be the case with the reporting of AIDS. Before 1982, only 15 cases of AIDS had been officially reported to WHO from countries in the European region. According to WHO statistics, these included the following countries and numbers of cases: Denmark, 3 cases; France, 7 cases; Switzerland, 3 cases; Spain, 1 case; and West Germany, 1 case. Even by the end of 1982, the number of AIDS cases had increased more than fourfold to 69. By the end of 1984, reports on 562 cases had been received (Mann, 1987) and over 30,000 cases of AIDS had been reported from Western European countries by early 1990 (WHO, 1990). This represents 99% of the total number of cases—28,247—from the entire region.

Twenty-one cases of AIDS were reported from Belgium in 1982, the majority of which had ongoing strong links to and contacts with central African countries dating back to the 1970s. This observation, coupled with the now-recognized long incubation period of AIDS, suggests that the virus was well established before official reports of AIDS cases were made to WHO. The information contained in Table 5.1 helps in understanding sequential increases in the reporting of AIDS during the infusion stage in Europe.

According to retrospective searches by the Institute of Cancer Research (1984), at least 8 cases of AIDS had occurred in Europe prior to 1979, and 104 by 1983. As expected, less than half had been reported to WHO. In spite of underreporting, France appears to have been one of the first countries to be seeded with the HIV. In 1983, a French research group studying a sample of 27 AIDS cases found that all but two were from the Paris area (Bouvet, 1983). Sixteen were homosexual males, seven were heterosexual men, and four were women. Several had visited the United States and several others Haiti (three were Haitian). Others had visited countries in equatorial Africa and two were originally from Zaire.

TABLE 5.1. Reporting of AIDS Cases Pre-1982-1984

Country	Before 1982	1982	1983	1984
France	7	20	57	176
Switzerland	3	4	11	22
Denmark	3	4	12	17
West Germany	1	11	37	103
Spain	1	2	10	24
Belgium	—	21	33	49
Netherlands	—	3	15	30
United Kingdom	—	3	26	77
Italy	—	1	4	23
Greece	—	—	1	5
Austria	—	—	—	13
Sweden	—	—	4	12
Norway	—	—	—	5
Finland	—	—	—	5
Portugal	—	—	1	1

Source: Extracted from the World Health Statistics Quarterly, 40 (1987).

In contrast, almost 90% of 40 AIDS patients identified in Belgium in 1983 were of African origin, were not homosexuals, and were not intravenous drug abusers (Clumeck, 1984). Of the 41 AIDS cases identified in West Germany by September of the same year, 36 or 88% were homosexual males. Further, all of the German AIDS patients had traveled to the United States, Haiti, or central Africa from 1978 through 1983 (L'age-Stehr, 1984). Similarly, a 1983 study of 24 British AIDS cases indicated that one had been diagnosed in 1980, two in 1981, and 10 in 1982. Of the total number of persons with AIDs, 20 were homosexual males and most had had New World sexual contacts (McEvoy, 1984). Dutch researchers studying eight AIDS cases in the Netherlands in 1983 found that seven were homosexual men and four of these men had also had gonorrhea and syphilis (Vandenbroucke-Grauls & Verhoef, 1984). Half of the Dutch AIDS patients had had contact with male homosexuals from the United States. These early patterns of contact with the United States and central Africa reported in European studies of the early 1980s reinforce the notion that the disease did not originate in Europe.

The earlier European studies also indicate the extent to which HIV infection and AIDS were primarily urban phenomena during the infusion phase (Brunet & Ancelle, 1985). As shown in Figure 5.1, the distribution of AIDS cases in selected European cities in 1983 was heavily concentrated in large cities with major international connections. For example, 30% of all West German cases were located in Berlin and 20% in Frankfurt. Similarly, well over half of all cases in Switzerland were identified in Zurich (41%) and Geneva (24%). Italian cases reflected a similar pattern with 29% of the reported cases in Milan and 21% in Rome.

Other European AIDS core countries had even heavier concentrations in large cities. Paris claimed 90% of all French cases in 1983, and Madrid 83% of those in Spain. In addition, 76% of all Danish cases were located in Copenhagen, 73% of all reported British cases were in the London area, and 66% of Dutch AIDS cases were reported from Amsterdam. Thus it appears the infusion of HIV infection into Europe followed the same kinds

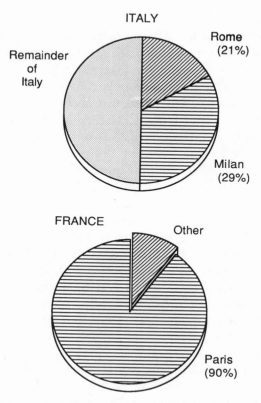

FIGURE 5.1. Concentrations of European AIDS cases in limited numbers of large cities were common during the early 1980s. These patterns reinforce the argument for hierarchical diffusion during the seeding phase and strongly suggest that the disease did not originate in Europe.

of spatial temporal diffusion patterns identified a few years earlier among major metropolitan centers in the United States.

THE PRIME SURGE IN EUROPE

By 1986, the geographical diffusion of AIDs in Europe had accelerated. As indicated in Figure 5.2, during the prime surge

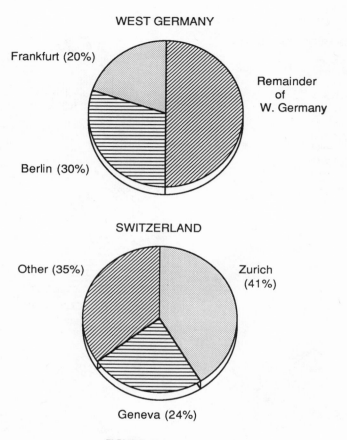

WEST GERMANY

Frankfurt (20%)

Remainder
of
W. Germany

Berlin (30%)

SWITZERLAND

Other (35%)

Zurich
(41%)

Geneva (24%)

FIGURE 5.1 *cont.*

phase the disease had spread from the Western European core countries as well as those countries comprising the core-fringe areas into peripheral countries. Based on official reports to WHO, the pattern actually represents conservative estimates of the diffusion of the HIV and associated cases of AIDS. The tremendous spatial and temporal disparity between Western and Eastern European countries suggests several possible explanations of the geography of AIDs in this part of the world.

One interpretation of the pattern is based on the previous political pattern of the region, in particular, the presence of the

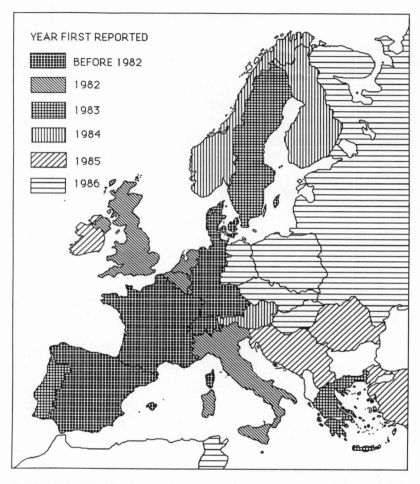

FIGURE 5.2. AIDS had spread from Western to Eastern Europe by the mid 1980s. This map shows general patterns of spread as expressed by year first reported (World Health Organization data).

so-called iron curtain. It may be surmised that this barrier to communication between the West and East initially prevented and later limited or retarded the west-to-east spread of the HIV and AIDS that might otherwise have been expected to occur. By early 1990, the eastern European countries of Bulgaria, Czechoslovakia, East Germany, Hungary, Poland, Romania (prior to the identification of the pediatric epidemic in 1990), Yugoslavia, and the USSR accounted for 228 cases, or less than 1%, of the total AIDS cases reported to WHO for all of Europe. Albania has reported no AIDS cases.

A possible second explanation for the differential distribution of AIDS cases between Eastern and Western European countries centers on the assumption that AIDS cases were present in the former but were underreported or not officially recognized or reported at all for several years (World Press Review, 1986). Recent reports from Romania provide some support for this notion (Bohlen, 1990).

In Romania the first documented case of AIDS occurred in a bar worker who was hospitalized in the Victor Babes Hospital in Bucharest in 1985. Romania reported about a dozen cases of AIDS by the end of 1989. However, evidence of AIDS among Romanian children is only surfacing now, and the story is dramatic. Apparently, the practice of injecting small amounts of blood into the umbilical cord of small babies in order to stimulate growth—a practice rarely reported elsewhere in the world in recent history—was widespread in Romania. This practice, coupled with the lack of blood screening for the HIV and the repeated use of syringes for injecting the blood, has resulted in a major epidemic of pediatric AIDS. An emergency team of doctors from WHO has already identified 700 children with AIDS. But the full extent of the situation will not be realized for several months, as the number of cases is reportedly especially high in orphanages located throughout the country from which accurate counts have not been received (Hilts, 1990). It is to be hoped that this situation is unique and that similar situations will not be discovered as other repressive governments are dismantled in Eastern Europe and elsewhere. Official underestimates of the disease clearly have not been restricted to Eastern Europe.

Yet another explanation of low reported numbers of persons with AIDS suggests that in spite of lags in reporting and variable official attitudes and behaviors toward the disease, it actually did spread more slowly from west to east as a function of increased distances from points of origin and levels of urbanization.

As already indicated, France, Denmark, Belgium, the United Kingdom, Switzerland, and West Germany were probably seeded before 1980. These countries subsequently moved through the infusion phase of the epidemic by 1984, and served as Europe's core countries for AIDS diffusion. Temporal increases in AIDS cases for these countries (except the United Kingdom) are shown in Figure 5.3. By the end of 1986, France had more reported cases of AIDS—1,200—than any other Western European country. France was followed by West Germany, which had reported more than 800 cases, and the United Kingdom, reporting more than 600. While cumulative numbers of AIDS cases were less for Denmark (131), Switzerland (192), and

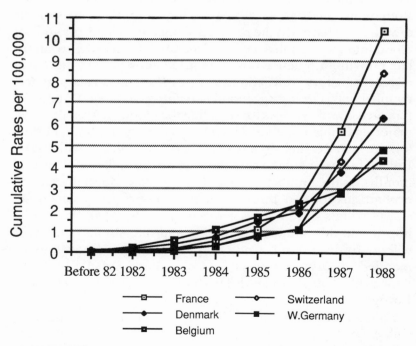

FIGURE 5.3. The temporal progression of AIDS in five European core countries during the 1980s.

Belgium (207), the per capita rates were fairly high by 1986 European standards due to their small national populations (WHO, 1987a). Even after 1986 per capita AIDS cases continued to accelerate at an almost exponential rate in some of the larger core countries and some of the core-fringe countries, for example, Spain, the Netherlands, and Italy. Again, such trends are characteristic of the prime surge phase of epidemics.

Yearly cumulative case records from the United Kingdom for the period from 1982 to 1989 exemplify the temporal increase trends that have been observed repeatedly in Western European countries. By the end of 1986, 610 cases of AIDS had been reported in the U.K. and 56% (339 cases) were new in 1986. Table 5.2 provides the numbers of new cases for each year of this 5-year period (note the characteristic temporal diffusion curve). These figures assist in understanding the importance of the doubling of cases. For example, there was almost a ninefold increase in the number of new AIDS cases reported between 1982 and 1983; this increase was reduced to a tripling between 1983 and 1984 and a doubling between 1984 and 1985 and between 1985 and 1986. In other words, beginning in 1984 the number of reported AIDS cases doubled approximately every 12 months. By the spring of 1989, it was estimated that the doubling time for new cases of AIDS in the United Kingdom had

TABLE 5.2. New Cases of AIDS, U.K., 1982-89

Year	New Cases
1982	3
1983	26
1984	77
1985	165
1986	339
1987	617
1988	755
1989	848

Source: World Health Organization, Weekly Epidemiological Report, February 6, 1987; July 31, 1987; January 8, 1988; March 2, 1990.

been extended from 12 to 18 months (BMJ, 1989). This was reportedly due to a leveling off of new cases among homosexual and bisexual men, and little penetration of the HIV into the heterosexual community. However, an increase in the number of new cases was expected from intravenous drug abusers in Scotland, many of whom became infected in 1983–84.

Transmission characteristics of the cases reported to WHO from the United Kingdom were different to a certain extent from those AIDS cases confirmed in the United States during the same period. The main similarity was the high percentage of new cases among homosexual and bisexual males (88% in the United Kingdom). However, only 4% of the cases were intravenous drug abusers. In the United States the percentage of AIDS cases among all intravenous drug abusers was approximately 17%.

There had been periodic indications of AIDS cases in Eastern Europe since the early 1980s, but it was not until 1985 that WHO began receiving official statistics and reports. In 1985 Yugoslavia reported two cases and Romania one case (Mann, 1987). By 1986 reports of AIDS cases were coming from Poland (1), East Germany (1), Czechoslovakia (6), Hungary (1), and the Soviet Union (13).

By 1987 Budapest had become an important Eastern European center for AIDS research and information (Glenny, 1987). Reportedly, a 52-year-old man was the first to die of AIDS in this country (in 1987). Compulsory testing was initiated at this time and 74 of the 107 persons registered as HIV positive in 1987 were homosexual men.

The relatively large number of cases in the Soviet Union can be accounted for by outbreaks of HIV infection in several hospitals. Here the penetration and diffusion of HIV infection can be traced to one homosexual male believed to have been infected in East Africa in 1982. He subsequently infected 5 of his 22 sexual partners in the USSR who, in turn, transferred the infection to three women in heterosexual intercourse (Pokrovsky, Yankina, & Pokrovsky, 1987). One of these women gave birth to a seropositive child.

As a result of blood transfusions from a donor who was infected with HIV via homosexual contact, five blood recipients became infected (Pokrovsky et al., 1986). In 1984 an eleven-year-old child was infected with HIV through a blood transfusion. Apparently against hospital regulations, common syringes were used and subsequently the HIV was spread to a number of children. By 1987, the entire medical system of the USSR had switched to disposable syringes (Rich, 1987). Additionally, anonymous HIV screening was established in Moscow; an important action because, under other circumstances, a 5-year prison term accompanies conviction of homosexual activity in the Soviet Union. Recent additional actions include screening of blood donors for HIV and interviews with Soviet citizens believed to be likely to have sexual contacts with foreigners.

Actually, mandatory HIV testing had already begun in 1986 in parts of Western Europe, and this practice no doubt spread eastward with the disease. In southern Austria, the Carinthian capital of Klagenfurt was reportedly the first Western European city to introduce compulsory HIV testing (Anonymous, 1987a). Klagenfurt is apparently "not known for its tolerance towards minorities"; thus black nightclub entertainers, some of whom make extra money as prostitutes, and foreigners working in Klagenfurt were some of the first people to be tested. By 1987 Bavaria in West Germany had also introduced mandatory HIV testing (Dechau, 1987), and police were instructed to enforce a statute that included testing among male and female prostitutes, prisoners, those held for ransom, and anyone else suspected of harboring the disease. Federal health authorities did not necessarily agree with the Bavarian policy, but similar sentiments in that part of Europe and in many other parts of the world had been expressed.

The HIV infection probably spread into Eastern Europe via tourism. It is doubtful that the disease would be spread to any great extent via intravenous drug abuse. Still, the disease has now diffused throughout Eastern Europe. Bulgaria reported its first case in 1987, but no cases have yet been reported from Albania, although WHO officials do suspect that the disease is present in the Albanian population, particularly since AIDS cases have been reported from the neighboring countries of Greece since 1983 and Yugoslavia since 1985. However, between August and December 1988, about 1,600 Albanians supposedly at very high risk for HIV infection were tested for HIV seropositivity using the most up-to-date analyses available, and reportedly, none of the subjects' blood samples tested positive for the HIV antibody (WHO, 1989c). If accurate, Albania represents a remarkably interesting "aberration" in the observed as well as the expected European diffusion pattern, perhaps due to its political and social isolation.

The magnitude of the increases in AIDS cases in Europe during its prime surge phase was generally well understood by early 1987. The number of reported AIDS cases had increased from 4,549 at the end of December 1986 to 5,687 by the end of March 1987, an increase of 1,138 cases or 25% (WHO, 1987b). During this quarter, the largest numbers of new cases occurred in France (411), West Germany (173), Italy (141), the United Kingdom (119), and Spain (93). Of the total number of new AIDS patients, 5% were African and 3% American. Ninety percent of the persons with AIDS were males. Homosexual and bisexual males represented 64% of the total number of cases, and 16% were heterosexual intravenous drug abusers. About 20% of the cases of non-European origin had apparently been infected through heterosexual activity.

For the one-year period from December 1987 through December 1988, France again had the greatest increase in the number of reported AIDS cases: from approximately 2,500 to 4,900, almost doubling within the year. During the late 1980s, the disease had also spread into southern France. The Mediterranean coastal areas of France contained about 40% of the AIDS cases in that country, a condition that contrasted with the 90% of known cases that showed up in the Paris metropolitan area five years earlier (Jougla, Hatton, Michel, & Letoullec, 1989). The number of cases in West Germany increased from about 1,500 to 2,600, up 60% over the year. AIDS cases reported in Italy increased from about 1,100 to 2,600, more than doubling during the same period. The United Kingdom registered one of the lower increases in new AIDS cases between the end of 1987 and the end of 1988, from 1,100 to 1,900, an increase of "only" 73%.

During the late 1980s, European public reaction to AIDS resulted in a variety of governmental actions. Such developments as the Bavarian and Austrian compulsory HIV testing exemplify the social–political conflict that the disease has generated in many countries. By 1987, Great Britain began to selectively prevent AIDS patients from entering the United Kingdom (Anonymous, 1987b). The British action included an incident wherein a man infected with the HIV was turned back by customs officials at Gatwick International Airport. British immigration officers have the authority to refer anyone whom they consider "physically abnormal" to undergo medical screening. In West Germany several elected officials recommended registration of AIDS patients (Neffe, 1987).

For the most part, however, European public health officials seem opposed to strict identification and quarantine as measures to control the spread of the HIV. By all accounts such measures would be nonetheless futile at this stage of the epidemic. Thus major efforts are concentrated in the development of education programs with heavy emphasis on the use of condoms during sexual intercourse. In addition, Great Britain and France have recently increased funding for the development of a vaccine against HIV infection (Dickson, 1987b).

The major political changes that swept across Eastern Europe in the latter part of 1989 and that are continuing into the 1990s may have important implications for the further diffusion of HIV. HIV infection thus far has spread into Eastern Europe for the most part via tourism. Now, for example, hundreds of thousands of residents of East Germany can travel freely to West Germany, and vice versa. With the increase in tourism between East and West, we may expect an intensification of the current geographic trends of HIV exposure throughout Eastern Europe. These Eastern countries may move more quickly than expected into the prime surge phase of the epidemic of HIV infection and AIDS. What may distinguish the intensity of patterns of HIV infections and resulting AIDS cases are significant differences in the distribution and use of prostitutes, numbers of intravenous drug abusers, and the extent of promiscuous sexual practices among resident male homosexuals.

HIV INFECTION AND AIDS TO DATE

In summary, we present here sequential and integrated diffusion pattern for Eastern and Western Europe:

1. *Seeding phase.* The first and most obvious stage in the sequence is the movement of the infection from one large city to another. As indicated earlier in this chapter, during the initial seeding phase of the epidemic, HIV infection and AIDS can basically be considered an urban phenomenon.

2. *Infusion phase.* The next stage in the pattern of diffusion reflects the contagious nature of the disease: it is the disease's spread outward from the largest metropolitan centers to smaller cities and surrounding areas. Since so much of Western Europe is heavily urbanized and, by Western international standards, densely settled, high risk groups for HIV infection and AIDS are perhaps spatially juxtaposed to a greater extent than in many other parts of the world.

3. *Prime surge phase.* During the infusion and particularly the prime surge phases of the epidemic, the disease spreads into villages located in the interstitial countryside and into other less densely settled areas. The latter situation is demonstrated by adjusting for the number of cases by calculating rates per 100,000 persons for each country and examining the experiences of proximate countries.

Three major diffusion pathways (or traces) through the stages can be identified. The first, the Scandinavian tract, includes Denmark, Finland, Norway, and Sweden (see Figure 5.4). Denmark was one of the earlier European countries to be seeded with HIV infection. By 1983, as indicated, HIV infection appears to have been well established in Sweden. And for the mid-1980s, a classic and characteristic diffusion curve inclusive of Norway and Finland can be identified. The Scandinavian countries

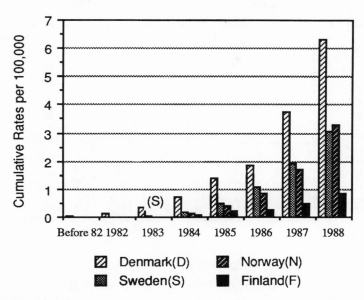

FIGURE 5.4. The temporal spread of AIDS along a postulated Scandinavian pathway from 1982 to 1988.

moved into the prime surge epidemic phase in 1987 and 1988.

The second pathway, the Central European diffusion tract, is also easily identified (see Figure 5.5). Reflecting the initial seeding, a comparison of the simultaneous temporal trends in Belgium, West Germany, Austria, and Hungary demonstrates their contiguity. Clearly the low Hungarian rates during the latter part of the 1980s can be attributed to isolating political conditions. Still, it is anticipated that Hungary will feel the prime surge of HIV infection and AIDS diffusion in the early 1990s, following Austria and West Germany by only a few years.

Thirdly, a Mediterranean diffusion tract of countries can also be identified. As shown in Figure 5.6, by 1986–87, a typical west-to-east European AIDS infusion curve could be identified in countries of the Mediterranean region. Bulgaria reported its first official case of AIDS in 1987, but apparently Albania's

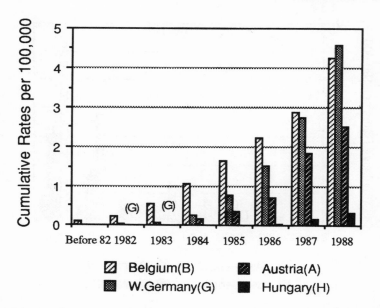

FIGURE 5.5. A probable Central European diffusion pathway during the 1980s.

political and social isolation from countries throughout the region has resulted in a retardation of the progress of HIV infection to this country.

EUROPE—TOWARD THE YEAR 2000

Most of the individuals who will develop AIDS during the next several years are already infected. Should a vaccine against HIV infection be forthcoming in the near future, we can expect a delay of several years from development to distribution. This situation, coupled with the long incubation period of AIDS, suggests that the outlook for HIV infection and AIDS in Europe during the next decade obviously may not be substantially altered even with the development of an effective vaccine against HIV infection in the near future. As mentioned above,

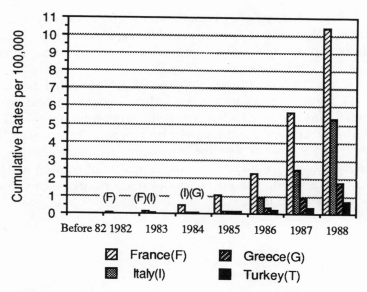

FIGURE 5.6. The progression of AIDS from the Eastern to Western Mediterranean is suggested from the graphic comparisons here.

spatial and temporal trends of the disease in Europe seem to be following those observed in the United States by several years.

While there are some striking similarities in AIDS trends in the two continents, there are some differences as well. In both regions there has been an increase in HIV infection and AIDS among women. In the United States in 1987, more than half of a sample of women who tested positive for AIDS were prostitutes and had abused intravenous drugs, and about one fourth were sexual partners of men who either had AIDS or were at high risk (CDC, 1987c). By the end of 1989 about 3.5% of the AIDS cases were women. In the United Kingdom, 4% of the cumulative total of persons with AIDS through March 1989 were women. Of these women 16% were intravenous drug abusers and 15% were heterosexual partners of men at high risk. More specifically, prostitutes tested in London, Paris, and Nuremburg in 1987 showed no signs of HIV seropositivity. However, female prostitutes in six other cities in West Germany, in Zurich, and in Pordenone, Italy, were HIV positive. In Athens, 6% of a group of 200 registered prostitutes tested for HIV were positive, and apparently none were intravenous drug abusers.

By 1989, the male-to-female ratio of HIV-infected persons was 17:1, but among intravenous drug abusers it had been reduced to as low as 3:1. Thus, while the preponderance of AIDS cases in Europe will continue to be homosexual/bisexual males from 20 to 49 years of age, the disease is expected to diffuse to younger age groups and increasing numbers of women will be at risk. As a result, the numbers of children infected by their mothers will also increase.

6

AIDS Patterns in the United States

In this chapter we focus our attention on the patterns of AIDS in the United States reflected by analyses of available statistics. Although the specific geographical origins of HIV infection remain unclear at this time, the actual geography of AIDS in the United States can be charted relatively accurately. However, such an observation must be qualified to pertain to a period beginning in 1984, when the Centers for Disease Control (CDC officially released state-based reports of AIDS cases. It can be reasonably argued that, during the 1980s, more public and medical attention was directed toward the HIV and AIDS than any other medical condition. In addition to the related social patterns of infection and disease, considerable attention was directed toward the spatiotemporal diffusion and distribution of AIDS.

Diffusion and distribution appear to have followed a standard pattern of diffusion subsequent to the introduction of HIV infection and associated AIDS cases into selected socioeconomic and "behavioral" groups in several of the largest cities in the United States. That is, over a relatively short period of time, the infection appears to have moved down the hierarchy of population centers to secondary and tertiary cities and ultimately to the small town and villages across the landscape. The pattern of diffusion is one of several possibilities, but it is predictable only within limited parameters at different geographical scales.

Although we acknowledge the lack of hard scientific evidence to support the various origin theories discussed earlier, it does appear plausible to suggest that HIV infection in the United States resulted from tourism and travel to Haiti and Europe. Regardless of its specific origin, the establishment and diffusion of the HIV in the United States appears to have the following sequence:

1. The introduction of the HIV by individuals via international air travel, and subsequently the first AIDS cases in the 1970s.
2. The development of clusters of HIV infection and AIDS cases in relatively narrowly defined neighborhoods in large cities.
3. Major outbreaks of ARCs and AIDS among homosexual/ bisexual males residing in or frequenting these early "incubator" neighborhoods of specific large cities such as New York, San Francisco, Los Angeles, Miami, and Houston.
4. The "spilling out" of the HIV epidemic, carried by high risk behavior travellers using both airline connections as well as major connector highways, leading to regional diffusion and more widely dispersed secondary diffusion nodes.
5. The formation of major AIDS core areas in the United States, including the cities mentioned above, as well as major interior metropolitan areas including Chicago and Denver.
6. The gradual interior expansion of the AIDS fringe and periphery zones that develop ahead of the major epidemic; and the subsequent shrinkage of the AIDS periphery areas becoming the more saturated fringe areas.

As previously mentioned, in its initial phases the pattern of HIV infection and AIDS diffusion in the United States reflected to a large extent the urban concentrations of populations at high risk. Apart from hemophiliacs infected with the HIV through tainted blood supplies, these populations included homosexual or bisexual males and intravenous drug abusers. Though considerably underreported, extensive information about U.S. AIDS cases is now readily available from the CDC. The numbers are aggregated by state and by metropolitan statistical area (MSA) in most reports from this agency. Based on earlier CDC data, other studies have described the general geographic progression of AIDS in the United States (Dutt et al., 1988; Gould, 1988; Shannon & Pyle, 1989). More complete and detailed data now permit a more comprehensive analysis of the diffusion patterns and, further, suggest that there are few geographic areas in which the infection does not exist, posing a potential threat to the general population.

GENERAL DIFFUSION PATTERNS— THE NATIONAL PICTURE

As mentioned earlier, geographical aspects of the early phases of the introduction of HIV into United States population are not clearly documented. Delays in making public the magnitude of the early AIDS situation apparently derived in part from scientific disagreements within the CDC pertaining to the nature of the infection and associated diseases. However, such delays by public health authorities are not uncommon throughout history, and developing a consensus of informed scientific and medical opinion takes considerable time. Nevertheless, reliable data for state totals, including summary tabulations, are available beginning with the period from 1981 to 1983. There are some interesting maps of early-1980s core areas produced from crude rates (Dutt et al., 1987). And we know now that the general assumption that the entire population of the United States was at risk during the early 1980s was inaccurate. Conversely, geographical risk patterns of the 1990s will be much more extensive than the previous decade.

In order to demonstrate statewide geography of AIDS, given the restrictive characteristics of early at-risk population, a measure was developed for the number of cases and the relative share of all U.S. AIDS cases. This measure is referred to here as the "AIDS quotient." It is derived for each state as follows:

$$\text{AIDS quotient} = \frac{\text{percentage of state population with AIDS}}{\text{percentage of United States population with AIDS}}$$

This measure is superior to the use of simple crude rates for the early years of the AIDS epidemic in that it actually uncovers concentrations otherwise masked by the crude rates.

Using this measure, the unfolding of the AIDS diffusion sequence in the United States can be seen in very general terms by examining Figures 6.1 to 6.6. The early 1980s were characterized by concentrations of reported AIDS cases in New York, California, Florida, Texas and Colorado. In reality, most of the New York State cases were reported from the New York City metropolitan complex. Similarly, a large share of the California cases occurred in the San Francisco and Los Angeles regions. Other early centers of AIDS cases developed in Denver, Colorado; Miami and Fort Lauderdale in Florida; and Houston, Texas.

The net effects of the early urban hierarchical diffusion during the late 1970s can thus be determined by examining state maps of data accumulated through 1983 and 1984 (Figures 6.1 and 6.2). By 1984, HIV infection and associated AIDS cases were occurring in relatively large numbers among several of the largest coastal urban areas and at least one major urban center in the interior of the country. In all likelihood—though these centers are all connected by interstate highways—travel by air and, therefore, airline network connections were involved in the early diffusion of the HIV and subsequent development of AIDS

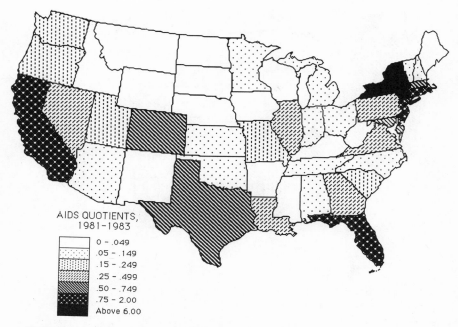

AIDS QUOTIENTS,
1981–1983

	0 – .049
	.05 – .149
	.15 – .249
	.25 – .499
	.50 – .749
	.75 – 2.00
	Above 6.00

FIGURE 6.1. Concentrations of AIDS cases in the U.S. based on first public releases of data (Source: Centers for Disease Control).

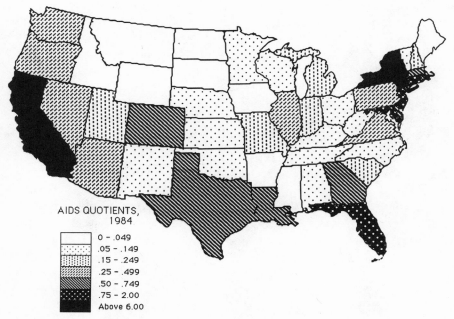

AIDS QUOTIENTS,
1984

	0 – .049
	.05 – .149
	.15 – .249
	.25 – .499
	.50 – .749
	.75 – 2.00
	Above 6.00

FIGURE 6.2. AIDS quotients determined cumulatively by the end of 1984. Increases could already be noted from core states.

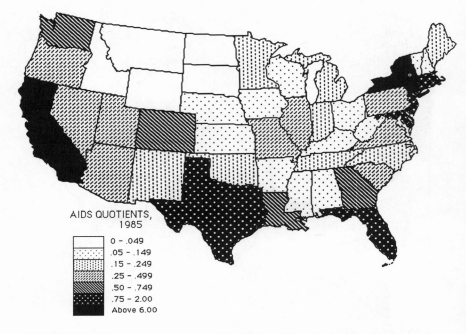

AIDS QUOTIENTS,
1985

- 0 – .049
- .05 – .149
- .15 – .249
- .25 – .499
- .50 – .749
- .75 – 2.00
- Above 6.00

FIGURE 6.3. This map depicts the spread of AIDS by the end of 1985.

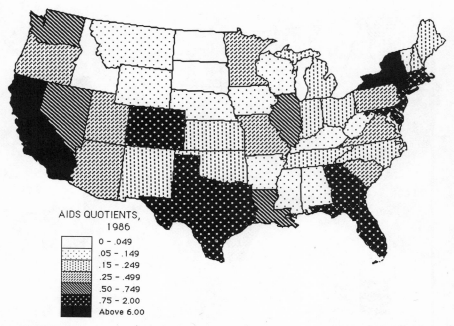

AIDS QUOTIENTS,
1986

- 0 – .049
- .05 – .149
- .15 – .249
- .25 – .499
- .50 – .749
- .75 – 2.00
- Above 6.00

FIGURE 6.4. The continued diffusion of AIDS as determined by AIDS quotients.

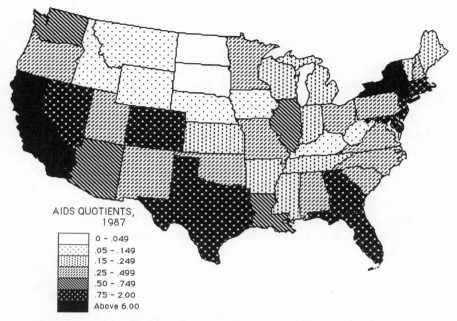

AIDS QUOTIENTS,
1987

	0 - .049
	.05 - .149
	.15 - .249
	.25 - .499
	.50 - .749
	.75 - 2.00
	Above 6.00

FIGURE 6.5. By the end of 1987, strong regional AIDS diffusion patterns had emerged.

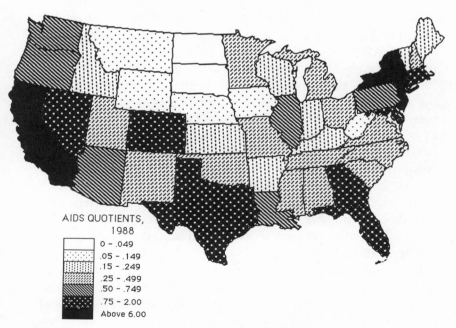

AIDS QUOTIENTS,
1988

	0 - .049
	.05 - .149
	.15 - .249
	.25 - .499
	.50 - .749
	.75 - 2.00
	Above 6.00

FIGURE 6.6. Coastal and interior contrasts in the spread of AIDS continued into the late 1980s.

in not only large metropolitan centers, but also major resort areas. The number and frequency of regular airline flights to and from certain destinations help explain the diffusion of AIDS during the early 1980s.

The general pattern developed in the early 1980s is reflected in the activities of a single individual, a homosexual male airline steward, designated as "Patient Zero" for the United States. A significant number of individuals among the first clusters of AIDS cases reported in Los Angeles, San Francisco, and New York were identified as having had sexual relationships with this individual or other individuals who previously had sex with him. The emergence of a substantial Gulf Coast pattern is most likely associated with the previously described travel of homosexual males from the United States to Haiti—as Patient Zero had on occasion—and the migration of infected Haitians to the United States.

The variable yet predominantly slower incubation period of the HIV compared to "fast" viruses, such as those associated with influenza, permits easier identification of the probable diffusion pathways of AIDS compared to some other diseases. For example, Figures 6.4, 6.5, and 6.6 illustrate the probable diffusion of AIDS beyond the initial parameters of major metropolitan areas along major interstate highway networks. Clearly, secondary patterns of regional diffusion had developed by the mid-1980s. The expanded national core areas of HIV diffusions at that time were California–Nevada, Colorado; Texas; Florida–Georgia, the northeastern corridor from Boston to Washington DC; and eventually the greater Chicago–northwestern Indian region. By 1986 core areas were thus both coastal and "interior-nodal." The national pattern of AIDS diffusion in the late 1980s appears to have followed directions already established during the previous years and determined by travel within the spatial network of the United States' air and highway transportation system.

It is notable that large areas of the "interior" United States initially did not experience substantial numbers of AIDS cases. As previously mentioned, Chicago eventually became a regional diffusion node, but there were proportionately fewer cases than would be expected based on the size of its population. The same was true for the Detroit metropolitan region, the Minneapolis–St. Paul area, and most of the area considered to comprise the traditional Midwest region.

The number of reported AIDS cases continued to increase during the mid-1980s, and a major revision in criteria used to determine the disease also led to a significant increase in the number of diagnosed cases. The CDC revised the AIDS case definition in September 1987 for persons with laboratory evidence of HIV infection. The revised definition included a broader spectrum of diseases characteristically found in persons with HIV infection. These diseases included a presumptive diagnosis of *pneumocystis carinii* pneumonia (PCP), HIV wasting syndrome, HIV dementia, esophageal candidiasis, and extrapulmonary tuberculosis (CDC, 1987a).

Between September 1987 and December 1988, of the 40,836 new cases reported, 11,966 (29%) met only the 1987 criteria. An estimated 20,000 to 30,000 additional cases were recognized within a few years subsequent to the revision. By the end of 1988, the CDC reported 82,764 cases of AIDS (CDC, 1989a). Of these cases, 93% were adults, and 91% of the adult cases were males aged 13 years of age and older. The proportion of female adult cases has gradually increased, from 7% before 1984 to nearly 10% by 1988. By the end of 1988, 3,589 people with AIDS were reportedly heterosexual and were infected by sexual contact with an individual at high risk for HIV infection. Moreover, 35% of the AIDS cases were reported in persons born in countries (mostly Haiti) with a high incidence of heterosexually acquired HIV. Throughout the 1980s, the racial composition of persons with AIDS remained basically the same. The case rates for AIDS was highest among blacks and Hispanics. During the same period, the proportion of AIDS cases attributed to intravenous drug abuse gradually increased. Prior to 1985, about 15% of reported AIDS cases could be traced to intravenous drug abuse. That share had increased to 20% by the end of 1988.

In retrospect, today it is possible to identify AIDS fringe and peripheral areas that had developed by the end of the 1980s, as well as the more easily determined core areas. For some of the reasons previously mentioned in this work, the infection and associated diseases continued to diffuse both nationally and regionally. The patterns identified in Figure 6.7 are the result of the diffusion that had occurred by 1988. Fringe areas or zones continued to move ahead of the expanding core areas until the AIDS periphery of the United States included several interior areas of the country. Many of these areas are more rural and some are relatively uninhabited. Furthermore, large parts of the AIDS periphery are not normally visited by persons generally considered to be at high risk for HIV infection. In fact, these areas of the country are characterized by few large cities or luxury resort complexes when compared to the core and fringe areas.

Even now, the AIDS periphery continues to shrink. Examination of reported cases for the period from 1988 to 1990 reveals that the disease has diffused into increasingly more remote sections of the country. The shrinkage patterns shown in Figure 6.8 indicate that, by the beginning of 1990, the areas of the United

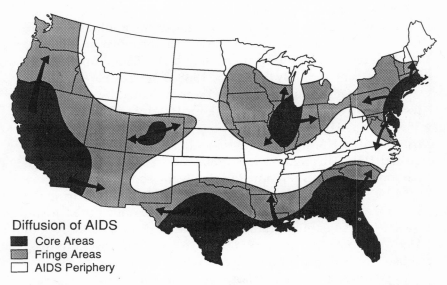

Diffusion of AIDS
■ Core Areas
▨ Fringe Areas
□ AIDS Periphery

FIGURE 6.7. AIDS core fringe and periphery areas had emerged by the late 1980s.

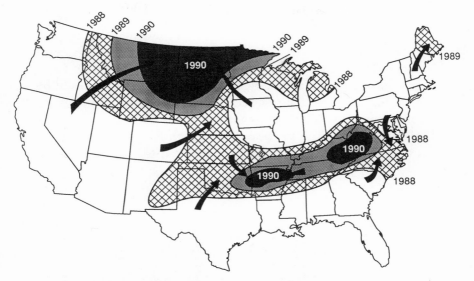

FIGURE 6.8. This map shows how the AIDS periphery had essentially collapsed by 1990 as the disease continued to spread into the interior.

States with the lowest AIDS rates were the Appalachians, the Ozarks, and the northern Great Plains. Given the fact that the incubation period of the HIV is very slow and that persons already infected with the HIV may not yet manifest disease symptoms or other illness conditions, even these remote areas will probably soon be included in the AIDS fringe regions as patterns of diffusion continue to be more localized.

REGIONAL ASPECTS OF THE DIFFUSION PATTERNS

As the AIDS fringe expands the size of the periphery is reduced and regional diffusion patterns become more pronounced. The most efficient method of identifying concentrations of AIDS cases below the state level derives from data on MSAs released by the CDC beginning in 1988. As more and more cases of AIDS are identified, regional concentrations can be determined by examining proportions of state cases contained within various MSAs. Figure 6.9 contains such selected metropolitan proportions for California, an early AIDS core state, and Ohio, until recently a state located in the AIDS fringe area.

In Ohio the proportions of AIDS cases in MSAs reported by the beginning of 1990 largely coincided with proportions of the population of the state residing in the MSAs. However, the comparison is not entirely accurate because most rural areas of the state still have lower AIDS case rates than the larger urban cen-

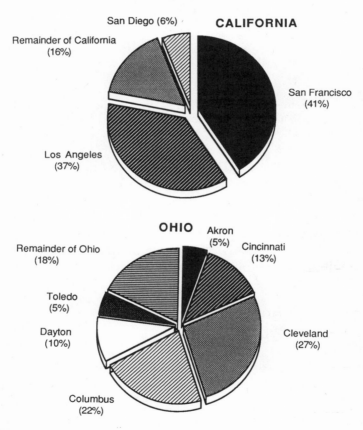

FIGURE 6.9. Metropolitan breakouts for California and Ohio by the beginning of 1990.

ters. Clearly AIDS is still an urban disease slowly diffusing outward and down the urban hierarchy in Ohio.

Regional increases outside metropolitan areas can also be noted for California, especially during the latter part of the 1980s. However, there are such relatively large numbers of cases in Los Angeles and San Francisco, two of the original AIDS core cities, that only two cities, San Diego and Oakland, contain as much as 5% each of the total number of reported AIDS cases in the state. At the same time, Anaheim contained about 4% of California's AIDS cases. Yet, at the beginning of 1990, Los Angeles accounted for 35% and San Francisco 32% of the total number of AIDS cases.

A comparison of similar MSA–state proportions of AIDS cases for Texas and Florida (see Figure 6.10) suggests that the disease is more widespread in Florida. For example, the Miami metropolitan area contains about 30% of Florida's AIDS cases while another 16% can be found in Fort Lauderdale. These two urban centers were the first to report significant numbers of AIDS cases during the early reporting period. By 1990, the West Palm Beach and Tampa areas each contained about 11% of Florida's cases, while Jacksonville and Orlando each accounted for another 5%. In contrast, by the end of the 1980s, Houston still remained the center of AIDS in Texas with 43% of the reported cases. Dallas was the other major center, accounting for 24% of the reported cases. Other Texas cities with more than 5% each of the AIDS cases included San Antonio (7%), Austin (6%), and Fort Worth (5%). Recently, reported AIDS cases in El Paso have been increasing.

Regional patterns of reported AIDS cases in Texas and Florida are well documented in a 1989 study of HIV seropositivity among applicants for military service (Gardner et al., 1989). In this study Pensacola and Jacksonville were identified as secondary diffusion centers in Florida. Also, much of the more widespread diffusion of HIV infection and AIDS in Florida has been attributed to travel patterns associated with the highway network. For example, in analyses of Texas patterns, a strong regional corridor of AIDS was identified from Houston to Dallas and another extended from Houston to Austin. Evi-

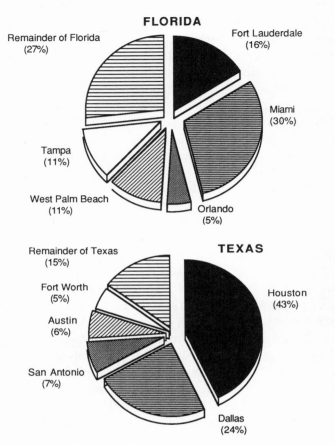

FIGURE 6.10. Metropolitan distributions for Florida and Texas by early 1990.

dence continues to point toward simultaneous intraregional and national diffusion. Regional distance–decay functions are manifest in the general decline in HIV infection away from the core areas in New York, California, and Texas. These distance-decay relationships can now be identified in many other sections of the United States.

Another method of determining the magnitude of the regional diffusion of AIDS is through a "numerical transect." With MSA

AIDS data it is possible to demonstrate probable paths of regional diffusion. Figure 6.11 presents west- to-east transects of the United States utilizing cumulative AIDS rates for selected MSAs for time periods ending in 1987, 1988, and 1989. It is apparent that San Francisco, Denver, and New York functioned as major national diffusion centers for the entire three-year period, and Chicago emerged as a node during 1988 and 1989. Regional patterns of diffusion are also clearly defined on a national basis as rates seem to both increase across the board each year and "abate" with increasing distance from major HIV-AIDS metropolitan centers. Again, similar distance–decay relationships had been noted earlier for San Francisco and New York (Gardner et al., 1989).

It is not surprising that the New York metropolitan area is the most complex with respect to HIV diffusion and associated distribution of AIDS cases—in part due to the sheer magnitude of the problem there. By the end of 1989, a cumulative total of nearly 25,000 reported cases of AIDS had been identified in the New York MSA. No other metropolitan area in the United States approaches this number. When cumulative rates per 100,000 persons are compared, however, the Newark MSA is a close second to New York (see Figure 6.12).

In spite of the magnitude of the problem in New York and Newark, a north–south transect of the original Boston-to-Norfolk megalopolis provides another indication of the nature of regional diffusion patterns within broadly inclusive urban agglomerations. There appear to be two major diffusion nodes along the eastern seaboard of the United States, namely, New York City and the District of Columbia. As proportionately more AIDS cases occur among intravenous drug abusers and prostitutes, control programs in these and other similar cities are essential. Meanwhile, the core area comprised of this megalopolis continues to expand, although proportionately fewer AIDS cases are reported from here over time.

It has become possible to reconstruct patterns of AIDS diffusion away from coastal areas. An examination of the Ohio experience for the period from 1981 to 1987 serves to illustrate how the problem developed in an initially low risk area of the United

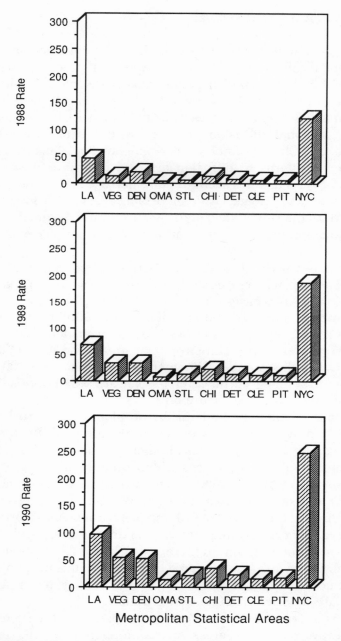

FIGURE 6.11. West-to-east cross-sections of increasing AIDS reporting for metropolitan areas for the 1988 to 1990 time frame.

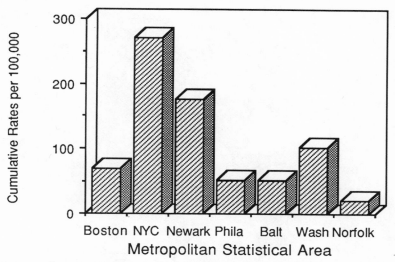

FIGURE 6.12. A north-to-south transect of megalopolis AIDS rates in 1990.

States, considered to be on the AIDS fringe (Gould et al., 1988). The first AIDS case reported for Ohio was located in the Cleveland MSA, the largest in the state. Interestingly, the second case was reported from Canton, part of an MSA situated about 50 miles south of Cleveland. By 1983, a hierarchical diffusion had taken place at several levels as the infection and associated cases of AIDS was reported in Columbus, Cincinnati, and Akron (Figure 6.13).

Within several years, there was clear evidence of radial diffusion as reported AIDS cases reflected a radially contagious pattern moving away from the major urban nodes as well as increasing within them. The northeast-to-southwest urban settlement pattern of Ohio was apparent in the AIDS distribution pattern that emerged by 1987. A comparison of reported AIDS cases and cumulative rates through the end of 1989 (CDC, 1989b) for major Ohio MSAs further indicates the presence of core and fringe areas within states. As indicated in Figure 6.13, the largest concentrations of AIDS cases remain in the more urbanized section of Ohio.

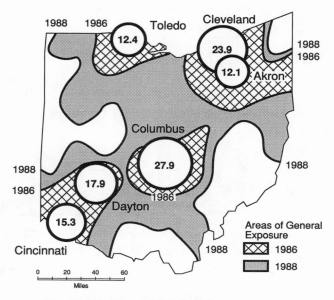

Notes: Circles include cumulative to 1990 rates per 100,000 persons.
Isolines indicate approximate "epidemic front" by 1986 and 1988.

FIGURE 6.13. The diffusion of AIDS in Ohio along with metropolitan rates. Circles include cumulative to 1990 rates per 100,000 persons. Isolines indicate approximate "epidemic front" by 1986 and 1988.

There are sections of the United States with only more recent experiences with AIDS that are now experiencing the impact of regional diffusion of the HIV. One such example is North Carolina. During the 1980s, North Carolina was geographically on the periphery in terms of the number of reported AIDS cases, but by the beginning of 1990, approximately 1200 AIDS cases had been reported. In 1988 and 1989 the state experienced a 67% increase in the number of reported AIDS cases. Developing trends indicate in the coming decade that North Carolina will reflect diffusion patterns similar to those that have already been identified in many other regions of the country. By January 1990, 58% of reported AIDS cases in North Carolina occurred in the three largest MSAs. The Raleigh–Durham MSA accounted for 22% and Charlotte accounted for an additional 21% (see Figure 6.14).

As HIV infection continues to diffuse nationally and regionally, additional urban centers in North Carolina and other fringe-area states (see Figure 6.7) are expected to report increased numbers of AIDS cases. A general southwest-to-northeast crescent of AIDS cases paralleling the general pattern of piedmont urbanization seems to be emerging. A broad metropolitan perspective assists in understanding the complexity of the problem in North Carolina as well as in other parts of the nation.

METROPOLITAN PERSPECTIVES ON AIDS

By the end of January 1989 over 120,000 cases of AIDS had been reported to the CDC. Of these over 100,000 were reported from MSAs. Cumulative rates for nearly 100 of these urban areas are listed in Table 6.1. Use of these tabulations permits an even more detailed spatial and temporal comparison. Broad AIDS diffusion "fields" can be identified for many parts of the MSAs in the United States.

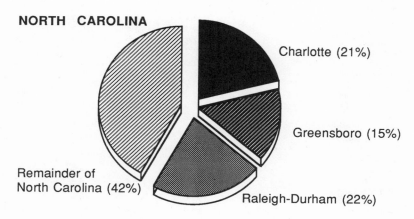

FIGURE 6.14. The metropolitan distribution of AIDS cases in North Carolina in 1990.

TABLE 6.1. Cumulative AIDS rates Per 100,000 by MSA to January 1, 1990

MSA of Residence	1/1/88	1/1/89	1/1/90	Total Cases
Akron, Ohio	4.3	8.5	12.1	78
Albany-Schenectady, NY	12.2	20.7	29	248
Allentown, Pa.	5.9	10.5	15.2	100
Anaheim, Calif.	18.8	29.7	42.9	930
Atlanta, Ga.	32.5	56.7	90.5	2316
Austin, Tex.	24.1	39	67.2	488
Bakersfield, Calif.	4.7	8.7	16.2	80
Baltimore, Md.	19.3	33	53.5	1220
Baton Rouge, La.	8.9	15.2	24.5	134
Bergen-Passaic, N.J.	37.3	60	75.8	985
Birmingham, Ala.	7.3	14.9	23	211
Boston, Mass.	29.3	49.5	70.2	1983
Bridgeport, Conn.	39.6	65.7	95.7	425
Buffalo, N.Y.	6	10.1	17.9	173
Charleston, S.C.	10	18.5	31.9	155
Charlotte, N.C.	7.2	13.6	21.9	233
Chicago, Ill.	18.1	32.2	47.1	2916
Cincinnati, Ohio	6.2	11.1	15.3	218
Cleveland, Ohio	10.1	17.1	23.9	443
Columbus, Ohio	10	19.4	27.9	363
Dallas, Tex.	37.3	60	82.5	1980
Dayton, Ohio	7	11.9	17.9	167
Denver, Colo.	26.1	41.8	61.5	1005
Detroit, Mich.	7.5	15	23.1	1002
El Paso, Tex.	4.5	7.4	13.3	75
Fort Lauderdale, Fla.	56.9	88.5	141.3	1614
Fort Worth, Tex.	12.4	22.2	34.4	431
Fresno, Calif.	7	13.8	22.3	131
Gary, Ind.	3.6	4.9	10.2	63
Grand Rapids, Mich.	3.5	6.8	10.9	71
Greensboro, N.C.	5.7	10.1	18.2	164
Greenville, S.C.	5	7.2	13.5	82
Harrisburg, Pa.	7.3	14.2	22.5	130
Hartford, Conn.	19.6	34.2	53.6	396
Honolulu, Hawaii	19.3	30	44.8	366
Houston, Tex.	50.3	78.6	106.2	3432
Indianapolis, Ind.	8.2	11.4	25.5	309
Jacksonville, Fla.	16.8	35.8	58.7	501
Jersey City, N.J.	108.9	193.5	249	1377
Kansas City, Mo.	13.9	29.9	45.8	696
Knoxville, Tenn.	5	9.8	13.7	81
Las Vegas, Nev.	18.2	33.5	55.6	317
Little Rock, Ark.	7.9	14.4	20.4	103
Los Angeles, Calif.	48.4	71.7	99.5	8256
Louisville, Ky.	4.6	9.2	13.8	133
Memphis, Tenn.	6.2	17	25.8	248
Miami, Fla.	78.5	118.3	169.2	2995
Middlesex, N.J.	27.2	48.6	68.3	649

cont.

TABLE 6.1. *cont.*

MSA of Residence	1/1/88	1/1/89	1/1/90	Total Cases
Milwaukee, Wis.	7.3	12.1	17.3	238
Minneapolis-Saint Paul, Minn.	10.4	16.9	23.7	544
Monmouth-Ocean City, N.J.	15.9	31	50.4	471
Nashville, Tenn.	4.5	18.7	27.1	252
Nassau-Suffolk, N.Y.	20.9	34.4	48.5	1277
New Haven, Conn.	38.9	67.6	89.25	457
New Orleans, La.	36.2	54.3	77.3	1032
New York, N.Y.	138.7	209.2	267.5	22665
Newark. N. J.	40	127.8	177.6	3354
Norfolk, Va.	8	14.9	20.6	269
Oakland, Calif.	33.5	52.2	72	1393
Oklahoma City, Okla.	10	13	14.9	147
Omaha, Neb.	4.9	10.7	14.7	90
Orlando, Fla.	15.5	35.7	53.7	482
Oxnard-Venture, Calif.	5.7	12.6	17.7	108
Philadelphia, Pa.	21.5	34.7	50.9	2455
Phoenix, Ariz.	13.6	25.3	37.8	719
Pittsburgh, Pa.	7.9	12.3	19.7	418
Portland, Oreg.	19.1	31.3	46.3	534
Providence, R.I.	17.7	30.1	43.1	273
Raleigh-Durham, N.C.	11.5	23.1	38.1	248
Richmond, Va.	12.1	20.1	31.6	256
Riverside-San Bernardino, Calif.	14.7	23.9	37.1	743
Rochester, N.Y.	10.9	20.8	27.3	268
Sacramento, Calif.	14.5	25.4	38.4	496
Saint Louis, Mo.	6.9	13.8	21.6	516
Salt Lake City, Utah	7.5	14.1	19.6	204
San Antonio, Tex.	10.8	29.3	44.4	567
San Diego, Calif.	31.2	51.4	74.3	1635
San Francisco, Calif.	240.6	352.5	465.1	7386
San Jose, Calif.	17.3	28.8	37.4	525
Scranton, Pa.	4.9	7.7	11.2	81
Seattle, Wash.	30.3	45.1	65.6	1149
Springfield, Mass.	8.9	13.7	22.2	115
Syracuse, N.Y.	8.5	13.1	19.1	124
Tacoma, Wash.	7.3	11.8	18.6	99
Tampa, Fla.	18.2	39.3	59.8	1144
Toledo, Ohio	4.6	8.7	12.4	76
Tucson, Ariz.	15.4	21.1	31.2	188
Tulsa, Okla.	6.3	12.8	19.5	143
Washington, D.C.	45.9	68.6	92.7	3303
West Palm Beach, Fla.	56.4	96.4	141.4	1069
Wilmington, Del.	11.4	20.9	32.5	179
Worcester, Mass.	10.3	21.1	30.9	126

Clearly, the greater New York City area stands out as the metropolitan area that had the greatest concentrations of reported AIDS cases in the 1980s. In spite of increasing proportions of AIDS cases occurring in other sections of the nation, both absolute numbers of AIDS cases as well as the cumulative case rates per 100,000 persons are notable. The cumulative number of AIDS cases in the New York MSA is about 23,000. This translates into a cumulative case rate of AIDS of approximately 270 cases per 100,000 people. The rates in adjacent New Jersey MSAs are almost as high, 250 per 100,000 (1400 cases) in the Jersey City MSA and 180 per 100,000 in Newark. Metropolitan rates do decline with distance from Manhattan, as the Nassau–Suffolk MSA has a rate of about 50 per 100,000—near the national average, while the Bergen–Passaic MSA has a rate of more than 75 AIDS cases per 100,000 people. Even Bridgeport Connecticut—still within the daily commuting field of New York City—has an AIDS rate nearly twice the national average, with more than 400 cumulative cases at the beginning of 1990.

In California, as one moves away from the San Francisco and Los Angeles nodes of AIDS concentration declining rates are also observed. With 7,300 cumulative reported cases of AIDS, the San Francisco MSA had a cumulative case rate above 450 per 100,000 persons at the end of 1989. Across the bay, the Oakland MSA rate was only about 70 per 100,000 and San Jose had a cumulative rate of only 35 per 100,000. The San Francisco "field" appears to have a much more steeply declining AIDS gradient than either New York or Los Angeles. This is due primarily to the presence of a large homosexual male community and the rapid course of infection among male homosexuals in the early stages of the epidemic. The cumulative case rate for Los Angeles at the end of 1989 was under 100 per 100,000 persons, still about twice the national average. The rate for Anaheim MSA, about 42 per 100,000, approaches the national average, while that for the Riverside–San Bernardino MSA is about 32 per 100,000. Both the major California AIDS core nodes of San Francisco and Los Angeles reflect general distance–decay patterns associated with the distribution of settlement densities.

Similar patterns also exist in MSAs of the southern United States. For example, the Miami MSA, long a major urban core area for AIDS, had a cumulative AIDS case rate of almost 165 per 100,000 people by the beginning of 1990. To the north, nearby Fort Lauderdale and West Palm Beach both reported case rates in excess of 135 per 100,000 persons. In the same manner New Orleans, with an AIDS case rate of 75 per 100,000 persons, is at about 1.5 times the national rate but over twice that of nearby Baton Rouge, 23 per 100,000. In Texas, Houston still has a cumulative case rate twice the national average, but the Houston MSAs rate has more than doubled since 1987. Dallas continues to demonstrate a high rate—77 per 100,000—while nearby Fort Worth has a rate of 35 per 100,000. As with so many others in the United States, these states continue to reflect the higher case rates occurring in the more highly urbanized areas. As more data becomes available, however, this pattern can also be examined more closely.

INTRAMETROPOLITAN PATTERNS

As already mentioned, large percentages of persons with HIV infection and AIDS developed very early in cities such as Los Angeles and New York City. We are now beginning to receive information on clusters of AIDS cases and HIV infection in large urban areas. For example, a recent map of the cumulative numbers of AIDS cases in metropolitan Los Angeles County depicted some 6,135 AIDS cases reported through the end of 1988 (Greenwald et al., 1989). The number of reported AIDS cases by residence at time of diagnosis was concentrated particularly northwest of downtown Los Angeles and in the communities of West Hollywood, Hollywood, Studio City, North Hollywood, Los Feliz, Silver Lake, and Glassel Park. A secondary focal point of AIDS cases appeared in Long Beach.

Similarly, a recent study estimates the geographic distribution of HIV infection within the Bronx borough of New York City (Drucker & Vermund, 1989). With a population of 1.16 million, it was estimated that there were just over 40,000 intravenous drug abusers in the late 1980s. Through January 1987, about 1% of the total cases of AIDS reported in New York City came from the Bronx. Among the six health districts in the Bronx, about two-thirds of AIDS deaths related to intravenous drug abuse occurred in only three South Bronx health districts.

These two studies indicate that we are now beginning to obtain information on a scale important to public-health policy development and program implementation. Moreover, this type of information at the intrametropolitan level permits the conceptualization and development of more specific spatial and temporal diffusion models of HIV infection.

CONCLUSION

More than half of all persons diagnosed with AIDS in the United States since 1981 have died. AIDS continues to increase annually in importance in terms of potential years of life lost, ranking sixth in both 1987 and 1988 (CDC, 1990a). In 1988 the total number of years lost was 472,800, about one-third the number of years lost to heart disease (about 1,500,000) and about one-quarter the number of years lost to malignant neoplasms (about 1,800,000). In part the large number of potential years of life lost is due to the relatively young ages of the majority of patients infected with the HIV and dying of AIDS.

It is also clear that female prostitutes in the United States are increasingly at risk to infection with the HIV (CDC, 1987b, 1989b). Among those tested, primarily those seen at clinics for sexually transmitted diseases, estimates now range from 13% to 22% infected. These figures may be distorted, however, due to the frequent intravenous drug abuse by prostitutes. Other individuals such as health care workers, physicians, and dentists are included in those occupations at an increased level of risk to HIV infection (CDC, 1987d, 1987e, 1988b).

The final configuration of the AIDS epidemic in the United States is still not clear, as many questions pertaining to its course remain. Of particular importance is the extent to which

HIV infection may be spread through the heterosexual population in the future. Data indicate that, though small, the percentage of females infected with the HIV and developing AIDS is gradually increasing. There is additional evidence, however, that the final form of the epidemic in the United States may be determined by ethnicity and socioeconomic status. Though the HIV infection is moving to fill the geographic, demographic, and socioeconomic interstices of the United States and its population, increasing percentages of HIV-infected individuals and persons with AIDS are poor and are members of ethnic minorities. Frequently, these groups are concentrated in certain segments of larger cities. Moreover, these people are the most difficult to reach with educational programs and, once infected, do not have adequate income or insurance to cover the extraordinary costs now associated with medical and palliative care. In the next chapter our attention is directed to some of the major implications of HIV infection and AIDS for the health care system of the United States.

7

AIDS in the United States: Implications for Health Care

The appearance of a new disease in epidemic proportions is likely to provoke severe anxiety in the general public and to create major problems for the health care system. AIDS has not only caused these problems, but has created unprecedented challenges to the health care system and to society at large because of certain unique attributes. On the most general level, the disease has reversed some of the recent gains in health status in the United States, especially among adult males in the more productive years of their lives. AIDS has tested the current arrangements for the delivery of health services to the population and has found them wanting. This epidemic has posed uniquely difficult challenges for health care systems here and elsewhere. Hence, we discuss some of the distinctive features of the disease and the attendant challenges facing the health care system. Our discussion focuses primarily on five such features.

THE POSSIBILITY OF A CURE

Perhaps the most significant attribute of this disease at the present time is its uncontrollability. We are painfully aware that there is no cure nor vaccine for AIDS. While the etiology of AIDS has been explained and the virus that causes the disease isolated, we are several years away from an effective cure or vaccine. This is not because of lack of concern or ineffective scientific work. Indeed, the speed with which the causes and methods of transmission of the epidemic were identified has been remarkable, given the complexity of the task. Nonetheless, scientists working on vaccines and remedies have estimated that it will take another 10 to 15 years before an effective cure or immunization is developed, some recent advances in treatment notwithstanding.

Because of the nature of the HIV and its infectious process, should a vaccine be developed, it will not be a panacea (Osborne, Jensen, Cooke, Koenig, & Lee, 1986). What this means is that the number of AIDS cases will continue to increase as we try to gather the armamentarium to stop the infection of new persons or to treat the persons who are already infected with the virus. In the United States, over 120,000 persons have been diagnosed with AIDS as of February 1990. As an indication of the magnitude of the problem in at least one geographic region of the United States, an estimated "two percent of Black pregnant women in New York State are HIV positive, about the same in Nairobi, Kenya, and about a third the rate in Kinshasa, Zaire" (Curran, 1989).

The estimated number of HIV-infected persons in the United States in 1992 ranges from 700,000 to 1.4 million. Although there is some disagreement regarding the exact magnitude of the epidemic, the implications are clear. The toll from this disease will increase dramatically before it is brought under control. In 1990 over 50,000 new AIDS cases are expected. This figure will increase to between 61,000 and 100,000 new cases during 1993 (CDC, 1990). From 1990 through 1993 a total of between 230,000 and 310,000 new cases of AIDS is anticipated in the United States. The only effective avenue for dealing with the problem, for the foreseeable future, is prevention at all levels: primary, secondary, and tertiary.

Primary prevention is achieved by preventing the spread of AIDS through avoidance of transmission of the HIV. Secondary prevention minimizes complications in infected persons through aggressive therapy. Tertiary prevention minimizes pain and suffering in persons with AIDS. The major challenge for the health care system is to develop effective prevention programs at each of these levels and to integrate them into the mainstream of the health care system.

THE NEED FOR PRIMARY PREVENTION

The second attribute of AIDS is closely related to the one just described, namely, its amenability or susceptibility to primary prevention. While authorities in the field may disagree regarding the amenability of this disease to primary prevention, few would disagree with the necessity of it. Not only is the alternative—treatment—costly and futile, but primary preventive methods can have several desirable spillover effects, including possible reductions in other sexually transmitted diseases, the birth of AIDS afflicted babies, the use of intravenous illicit drugs, and teenage pregnancy. This is not to imply that primary prevention is simple to achieve or even that we have found effective ways to do it. What is does imply is the importance of primary prevention and the fact that the solutions to the spread of AIDS have other solicitous social effects.

Where primary prevention can be most useful is in preventing the transmission of the virus through one or more of the following methods: the sharing of body fluids through sexual activity especially among homosexual and bisexual men, the use of contaminated needles by users of intravenous drugs, and currently to a lesser extent the receipt of transfusions of contaminated blood and blood products and in pregnancy from mother to child. While the urgency to find a cure and vaccine continues, the only effective approach to curbing the spread of AIDS is to prevent the transmission of the virus from one person to another (Osborne, 1986)

In the absence of major success in primary prevention thus far, solutions to the AIDS problem on the secondary and tertiary levels are sorely needed to help with the immediate problem. Again, should these solutions be achieved, they can have desirable effects in other parts of the health care system. If effective and appropriate models for home health care and community programs are developed for the treatment of AIDS patients, for example, these same models could be utilized in the treatment of elderly patients and others in need of long-term care.

THE RANGE OF PATHOLOGY

AIDS manifests itself in a variety of diseases or health problems, ultimately leading to death. In the United States the primary diseases include Kaposi's sarcoma (KS) (although there is now some doubt about the relationship between KS and AIDS), *pneumocystis carinii* pneumonia, and a variety of other opportunistic infections. In many instances these diseases are hard to avoid because the disease agents actually exist in or all around us, and under normal conditions they do not cause any problems. It is only when the immune system is seriously compromised that these diseases manifest. Essentially, the syndrome we call AIDS represents an extreme end of the clinical spectrum of disease.

In the early stages of the epidemic, after the onset of AIDS, life expectancy was about one and a half years, with a range of between 9 months and more than 18 months, depending on the specific health problems encountered, the condition of the host, and the aggressiveness of the therapy. With more aggressive therapies these estimates are being upgraded. However, the shortest life expectancies are associated with PCP, likely to be prevalent among intravenous drug users. High risk homosexual males who engage in anonymous sex with multiple partners also had relatively short survival rates. But there is some evidence that homosexuals have changed their sexual behaviors at least temporarily and risks have declined as a result. Also affecting current survival rates is the substantial reduction of very high risk groups through mortality; that is, many of them have already died. Because of new drug treatments such as 3'-azido-2',3'-dideoxythymidine (AZT) and aerosol pentamidine, the life expectancy of AIDS patients has been increasing. Recent estimates of life expectancy of AIDS patients indicate a doubling of the 18 months to three or more years. However, it has been suggested that this gain might be offset by additional diseases that result from the ineffectiveness of the immune system. Nonetheless, this is an untested hypothesis, and we are not certain what real long-term benefits (beyond two years) are accrued from aggressive therapy (Scitovsky, 1989).

From the perspective of the health care system, the diseases associated with AIDS have the following important characteristics: They require intensive treatment, especially during advanced stages, typically in a hospital setting, and for extended periods ranging from several months to two or more years; and they are invariably terminal. As we shall see later, the extended need of AIDS patients for in-hospital care, in the absence of alternative community resources, makes the treatment very costly.

THE ASYMPTOMATIC INCUBATION PERIOD

A fourth significant attribute of this disease is its infectiousness during asymptomatic stages. This means that individuals who harbor the HIV in their bodies but have no clinical symptoms of the disease can readily transmit it to others. In fact, AIDS has a rather long incubation period. Current estimates indicate that about 60% of persons who test positive for the HIV will develop the disease within 10 years. Earlier it was estimated that a smaller percentage of HIV seropositives would get AIDS. Opinion varies as to the stage in which the disease is most infectious. Some believe that infected people are most infectious during the asymptomatic stage rather than when the immune system has been so destroyed that it provides few cells for the virus to invade and replicate itself. Nonetheless, the length of the period during which persons are infected and can infect other persons has raised several fundamental issues that must be addressed. For instance,

- Should there be mandatory screening either for high risk individuals or upon marriage, employment, or immigration to the United States?
- Should we require certain categories of people (for instance, those who routinely have multiple sex partners) who are likely to expose others to the virus to be tested for HIV in order to limit the spread of the disease?
- What should be done if they are found positive?
- Should some form of quarantine be imposed on HIV seropositives?
- Should treatment be extended to asymptomatic HIV positives on a preventative basis?
- What are the social and economic implications of screening?

The health care system has to address these questions and provide satisfactory answers. In the absence of adequate response from the health care system, other institutions (such as the legal or penal system) may step in and fill the void.

A SOCIAL DISEASE OF SOCIETY

Because of its varied social, ethical, and legal implications and its variable effects on different segments of society, AIDS has been characterized as a disease of the social system (Willis, 1989). Not only does AIDS reflect social characteristics, as in promiscuous homosexual and heterosexual behavior, drug abuse, and prostitution, but its social impact has been felt differentially by segments of society. "AIDS is clearly affecting mortality, though differentially, by age, gender, racial, and geographic groups." Further, the effect of epidemics such as

AIDS "extends far beyond their medical and economic costs, shaping the nature of cultural and social life" (Willis, 1989).

While the views regarding the social manifestations of the disease might vary, it is certain that the way we deal with the problems will reflect the basic character of the social system in which we live. Whether we choose to blame the victims of AIDS or demonstrate compassion and understanding toward them is more a reflection of the character of society than any presumed behavioral "transgression" such persons may have committed.

AIDS SHARED ATTRIBUTES

Finally, a discussion of the unique attributes of AIDS will not fully explain the magnitude of the problem without some mention of its common traits. In fact, AIDS shares some important attributes with other diseases that have placed exceptional demands on the health care system and challenged its traditional role. For instance, people with AIDS exhibit clinical symptoms of diseases for which treatments, albeit not cures, are available. As in many other diseases, therapeutic requirements vary by the stage of disease and the availability of alternative sources of care. Essentially, AIDS patients need hospitalization during acute stages of illness and supportive and palliative therapy at all times. Also, as with many other diseases, all medical intervention can offer is palliative therapy, though substantial gains in life expectancy have been achieved through aggressive therapy. Palliative therapy is no less appropriate here than anywhere else. For instance, AZT treatment has been found effective, but the cost for the typical individual patient is prohibitive: $8,000 to over $10,000 per year, often borne out of pocket.

Because of their extended needs, many AIDS patients deplete their financial assets and exhaust their private health insurance benefits before they die. Similar to other clients in need of long-term care, many end up on public welfare, as custodians of the state, in the final stages of their illness. The only way to become entitled to public welfare is to be poor, by either starting out or becoming poor. Although we do not have a precise global estimate of the public share of the entire cost of AIDS, the available data suggest that no less than one-third of the total direct cost of AIDS (i.e., the cost of taking care of AIDS patients, exclusive of indirect costs that result from losses in productivity and premature death) is covered by public programs including Medicaid and Medicare, the bulk being assumed by Medicaid. Like other patients needing long-term care, AIDS patients must be concerned about the quality of services offered to them when supported by public funds (see Andrulis, Weslowski, & Gage, 1989).

DEMANDS ON THE HEALTH CARE SYSTEM

From a discussion of the basic attributes of AIDS, we now turn to a discussion of the implications of these factors for the health care system. In general, the impact of AIDS on the health care system has been profound. As indicated earlier, AIDS has tested the existing arrangements in health care delivery and found them wanting. One example is in the difficulty of getting and maintaining health insurance coverage—a problem not unique to persons with AIDS. A significant proportion of the U.S. population (now estimated in excess of 11%) lacks any private or public health care coverage. The number of persons with AIDS without adequate health care coverage is difficult to measure because of exclusions (services not covered), limitations (noncoverage for preexisting conditions and benefit ceilings), copayments and deductibles (the amount the client is responsible for even when services are covered by insurance), and, very importantly, the termination of benefits. Hence, several issues have yet to be resolved concerning health care benefits for persons with AIDS, including the type of care that is covered, how much of it is covered, what providers are eligible to provide the services, and what kinds of settings are particularly appropriate for care.

In the discussion that follows, we focus on three major issues facing persons with AIDS and the health care system, including the cost of care, incidence of cost (or who pays), and unresolved ethical and legal issues.

Cost of Care

The cost of care is a primary concern for anyone receiving health care. Rapid increases in the cost of care tend to outpace our ability to pay and have raised the concern that, with unchecked cost increases, we will soon get to the point of not being able to afford the system we have in place.

Fortunately, early estimates of the cost of care for AIDS patients, like the dire projections of the size of the AIDS population, were found to be too high. Current estimates have reported much lower figures. For example, claims data from the California Medicaid program (Medi-Cal) showed a decline in lifetime costs from $91,000 in 1985–86 to $70,000 in 1986–87 and $63,000 in 1987–88 (Scitovsky, 1989). This was achieved through reductions in the rate of hospitalization (both admissions and length of stay) and reductions in the costs of hospital care for AIDS patients by performing several services such as blood transfusions and intravenous therapy on an outpatient basis. Nevertheless, there remain very large variations in the estimates of the cost of AIDS that are not easy to explain. According to one source (Quesenberry et al., 1989), "lifetime costs per AIDS patient range from $23,000 to $168,000." Moreover, it has been pointed out that "much of the variability can be attributed to differences in estimates of the mean lifetime inpatient days accrued by AIDS patients." Also noted was that some lifetime cost estimates were too low due to differences in accounting procedures used. For example, some of the samples were limited to patients who have died during a given time period, "which generally results in a sample biased toward shorter lived individuals."

Nevertheless, it is clear that the direct cost of care for AIDS patients is declining and that the major factor accounting for the decline has been the decrease in hospitalization, both in number of admissions and length of stay in the hospital. Some communities, for example, several in the San Francisco Bay area, have developed strong community support and self-help organizations that have contributed not only to the lowering of the cost of care but also to a more humane treatment of persons with AIDS and the rendering of emotional support for terminally ill patients.

In view of the decentralization of data gathering activities and the variety of accounting procedures used for cost analysis of AIDS cases and the resulting lack of comparability in the existing data sources, it is impossible to give accurate general estimates of the cost of care for AIDS patients. Nonetheless we know it is considerable, and the range of estimates for lifetime costs given above are reasonable reflections of the range of costs.

Expenditures for AIDS patients from Medicaid for fiscal year 1989 were about $950 million, and from Medicare about $55 million. It is projected that approximately $3 billion will be assumed in 1994 by these two sources. Total expenditures for AIDS in 1991 are estimated to be between $4.5 and $8.5 billion (Smith, 1989). To put this in perspective, the estimate for 1991 would constitute between 1 and 2% of the total personal health care expenditures in the United States.

Regional variations in the direct cost of care for AIDS patients have been investigated (Andrulis, Weslowski, & Gage, 1989). The highest costs were found in the Northeast, which includes New York state (see Table 7.1). The lowest costs were in the

TABLE 7.1. Average Inpatient Costs per AIDS Patient per year (1986)

Region	Annual Cost
Northeast	$23,421
Midwest	19,530
South	16,902
West	14,858

West which includes California within which some progress has been made in the development of an effective community-based support system for AIDS patients that concerns itself with a broad set of issues ranging from the promotion of scientific research to finding a cure to assisting persons with AIDS in the latest information on treatment and other forms of support.

Who Pays

In the United States, the major single payer for in-hospital care for AIDS patients is Medicaid, whereas the major payer for out-patient care is the patient himself. A recent study of the cost of hospital-based care for AIDS patients (Andrulis et al., 1989) reported that Medicaid paid 44% of inpatient care and 31% of outpatient care, whereas patients themselves paid 23% of in-patient care costs and a high of 50% of outpatient care. The high percentage of AIDS patients receiving benefits from Medicaid is significant in several ways, most notably, (1) a larger proportion of AIDS patients are dying poor than are starting out poor, and (2) this is the only instance in which the proportion of males in this welfare program far exceeds the proportion of females.

Geographic analysis of sources of payment reveals substantial differences between the four major regions of the country (see Table 7.2). For example, in the Northeast and the West (including California) more than one-half of inpatient costs are paid by Medicaid (54 and 55%, respectively); the second source of payment in the Northeast is private insurance (accounting for 29% of inpatient costs) and in the West is self-pay (accounting for 23%). Private insurance covers 44% of the cost in the Midwest, but only 30% in the South, where nearly one-half, or 48%, of the costs are borne by the patients themselves.

The geographic distribution of outpatient costs follows yet a different pattern. Only in the Northeast does Medicaid pay more than one-half of such costs (58%). In the South, Medicaid pays only 19% of such costs, while patients pay 69%. The West is somewhat similar to the South, but on a smaller scale. In the West, Medicaid pays 27% of outpatient costs, and patients themselves pay 53%. The Midwest is basically in the middleground

TABLE 7.2. Percentage of AIDS Inpatient and Outpatient Costs by Region

	Overall	Northeast	Midwest	South	West
Inpatient					
Medicaid	44%	54%	35%	18%	55%
Private	29	29	44	30	19
Self Pay	23	11	17	48	23
Medicare	2	2	3	2	3
Prisoner	2	4	1	2	0
Outpatient					
Medicaid	31	58	46	19	27
Private	15	14	23	11	16
Self Pay	50	19	26	69	53
Medicare	3	2	4	1	4
Prisoner	1	7	1	0	0

Source: "The 1987 U.S. Hospital AIDS Survey" by D.P. Andrulis, V.B. Weslowski, and L.S. Gage, 1989, *Journal of the American Medical Association, 262,* pp. 784–794.

both geographically as well as behaviorally, where Medicaid pays 46% and patients pay 26%.

The geographic concentration of treatment facilities for AIDS is quite striking. It is estimated that only 10% of U.S. hospitals treated 58% of the patients with AIDS. This demonstrates the uneven burden that AIDS has placed on the health facilities in the country. However, as long as the geographic prevalence of AIDS remains uneven, this problem is likely to continue. There are indications that AIDS is becoming less concentrated in the first-wave cities that accounted for over one-half of all cases in 1988—now less than one-third.

ETHICAL AND LEGAL ISSUES

The most controversial issue surrounding the relationship been AIDS and the health care system relates to testing for the virus in the population at large. The ultimate justification for widespread testing rests on two principles: providing medical benefits to the client (if found to be seropositive) and preventing the further spread of the virus. Arguments against testing have been based on ethical (the right to privacy), economic (cost effectiveness of mass screening), and social (stigmatization of persons with positive results) concerns. While the call for early detection may appear benevolent—for the sake of helping those affected and protecting society at large, as can be the case for the early detection of other diseases—further scrutiny reveals a complex set of issues that may mitigate against the benefits and argue for a slow cautious approach.

There is little disagreement that early detection of asymptomatic AIDS can be a useful tool in restricting the transmission of the virus and in administering therapies that could have greater benefit at an early stage. However, this may not justify widespread or universal testing, such as making screening a requirement for employment, marriage, or immigration. Discrimination in employment on the basis of a potential health problem could impose an additional burden on individuals who test positive, many of whom will have no symptoms for a decade or more. Moreover, the virus cannot be spread through casual contact. Given the nature of the virus, the primary methods of transmission have been limited to intimate sexual contact and the sharing of body fluids. Premarital testing for the virus has been tried in Illinois with largely negative results, revealing the problems of widespread mandatory screening, especially of the general low prevalence populations. "[D]uring six months of testing, the program identified only eight seropositive individuals. The estimated cost was $2.5 million, or $312,000 per infected person" (Turnock & Kelly, 1989). From a social perspective, the problem is just as grave. Given the negative stereotyping and stigmatization of persons with AIDS, once a person is identified as having the virus, in many instances, he or she can expect social isolation and rejection.

Arguments for testing some population groups have also been made with some justification. There are certain public-health benefits to screening high risk individuals (Rhame & Maki, 1989). Such testing can help deter the infection of others and provide medical benefits to seropositives at an early stage, such as preventive tuberculosis therapy, pneumonia and influenza vaccine, AZT therapy, and prophylaxis against PCP. Rhame & Maki also suggested that early detection can provide candidates for research. One study, among others, reported a higher likelihood of terminating certain forms of high risk sexual behavior among homosexual men once they tested positive (van Griensven et al., 1988). Before we reach a decision regarding testing two questions must be answered: First, why do testing, and how will testing alter current preventive measures and second, how good is the test? (Weiss & Thier, 1988).

The final problem to be discussed here is the appropriate site of care for AIDS patients: Most agree on the desirability of a solution, but little progress has been made to achieve it. Because of the extended care needs of persons with AIDS and the importance of a supportive environment for palliative care, home health care and community programs have been suggested as crucial sites of care. These sites are widely encouraged: The Health Care Financing Administration, for example, has been encouraging states to request waivers to exempt payments for home and community care for Medicaid patients (Taravella, 1988). Others have expressed strong views favoring home and community care sites, stressing the importance of social support (Adelman, 1988), learning from the experiences of the elderly (Benjamin, 1988), compassionate care (Johnson, 1989), and appropriate sources of palliative therapy (Mansell, 1988).

It is apparent that the health care system has not yet fully adapted to the treatment of persons with AIDS. Before we can resolve this important and growing problem, the system must move to adopt universally acceptable modalities of treatment that (1) alleviate the pain and suffering of persons with AIDS while treating them as normal persons afflicted with a terminal illness, (2) minimize the cost to society without adverse effects on the patients themselves, and (3) can be readily integrated into mainstream medical care.

8

Modeling the Geography
of AIDS

Models have been developed in an attempt to predict the growth of the number of HIV-infected persons at international, national, and local levels (World Health Organization, 1989); Brookmeyer & Gail, 1986, 1988; Anderson, Medley, Blyth, & Johnson, 1990; Drucker & Vermund, 1989). In addition, models have been developed that attempt to replicate the transmission paths for the spread of AIDS between subgroups in a population (Knox, 1986; Anonymous, 1988b; Anderson, May & McLean, 1988). Relatively little work has been done in modeling the spatial dynamics of this pandemic. In this chapter we present and discuss attempts to model the geographic diffusion of HIV infection and the development of the AIDS pandemic.

MATHEMATICAL MODELS: AIDS, HIV, AND GEOGRAPHIC SPACE

Existing mathematical models intended to depict the spread of HIV infection and AIDS, mostly developed during the late 1980s from early 1980s data, generally fall into three broad categories (Gail & Brookmeyer, 1988). These groupings include (1) models based on empirical extrapolation, (2) procedures utilizing "back calculation", and (3) projections based on compartmental models. *Empirical extrapolation* is a method that involves the past reporting of AIDS cases; the fitting of trends, usually logistic in nature, to curvilinear frameworks; and extrapolation of these trends into the future. The *back calculation* method, espoused by Brookmeyer and Gail (1988) attempts to examine not only known AIDS cases, but also estimates of HIV infection in general populations. The third method, using *compartmental models*, is the most sophisticated modeling procedure of the three, but it is also the most abstract because it is based on estimates of both future HIV-infection probabilities as well as actual AIDS cases.

Most of the AIDS models developed and used by the CDC favor the first and second methods. Extensive estimates of populations at risk have been developed by nearly 100 experts according to the CDC (1987f). These estimates vary, of course, with the mathematical methods used, and, as a result, extrapolations are made from observed rates. On the basis of a National Academy of Sciences workshop held in 1987, a procedure was developed wherein the number of AIDS cases diagnosed each year are used to determine those who will be diagnosed the next year. The process assumes:

if $a(t)$ = number of AIDS cases diagnosed
during year t
$i(t)$ = number of newly infected in year t
$d(x)$ = proportion of infected persons
expected to develop AIDS after x years

then

$$a(t) = \sum_{z=1978}^{t} i(z) \cdot d(tz)$$

The number of AIDS cases, $a(t)$, was determined from reporting to CDC, but $d(x)$, the progression rate, had been based on HIV infection/AIDS rates among San Francisco homosexuals, with 95% confidence limits. Three different possible infection curves are considered:

logistic $l(t)$ $\qquad = l/\{1 + k \exp(-rt)\}$

log-logistic $l(t)$ $\qquad = l/\{1 + (rt)^k\}$

dampened exponential $l(t) = k \exp(rt^x)$

Even beginning with the same database (again, 1984, see Chapter 6), these models diverge rapidly. In general, one curve might work well in explaining the spread of AIDS via intravenous drug abuse but not so well with spread through transfusions. In addition, models based on homosexual HIV/AIDS rates in one city should not be extended to the population of an entire country. Another modeling complication results from the change in AIDS-case definition made by CDC in 1987 (CDC, 1987f).

The back calculation method is endorsed by Brookmeyer and Gail (1988) in more short-term modeling of AIDS cases. This procedure theoretically assumes no new AIDS cases after 1987 in a multinominal distribution, and seems to conform well to pre1987 data. In fact, as recently as 1989, CDC AIDS summary reports incorporated results from back calculation modeling using both pre1987 and current case definitions in attempts to extrapolate future AIDS cases (CDC, 1989a). As indicated by Brookmeyer and Gail, the pre1987 definition sets AIDS-case limits too low. Conversely, confidence limits for extrapolating future cases using post1987 case definitions increase drastically within a few years of the projection base.

Most of the statistical–epidemiological models now used in American public health give little consideration to spatial variations in HIV/AIDS rates. One exception is the work of Gonzales and Koch (1987) in an examination of a logistical model accounting for depletion of susceptibles. Aware of the possibility that exponential growth rates vary by country, Gonzales and Koch created a fictional country to develop a model that they subsequently tested with Canadian data. They generally concluded that exponential models fit known high risk groups and that epidemic slowdown can be identified sooner than might be expected. A similar conclusion has been reached by Golub and Gorr (1990) in an examination of Ohio AIDS reports. Golub and Gorr not only shed new light on less-than-exponential AIDS rate increases, they also demonstrate the importance of geographical simulations of AIDS diffusion.

An earlier analysis of AIDS diffusion in Ohio by Gould, Gorr, and Casetti (1988) also indicated both a spatial and temporal diffusion slowdown once larger urban places had reached certain levels. By modifying a standard gravity model into "multidimensional AIDS interaction space" the epidemic was simulated utilizing Cleveland, the state's largest city, as a major diffusion node. Another step was taken by applying "adaptive spatial filtering" to AIDS reports by counties. Ultimately populations of counties became key determinants in predicting the spread of AIDS, and several logistic curves were fit. The results included overprediction in some rural counties and underprediction in some more urban counties. Clearly, future studies of AIDS diffusion need to include geographic modeling procedures.

The epidemiological back calculation method seems to hold the most promise, and the keys to such a modeling process of HIV-infection-rate estimates and known AIDS cases, as previously mentioned. Such a modeling procedure is incomplete, however, without the incorporation of distances (sometimes metric and sometimes nonmetric) between diffusion nodes and reliable estimates of populations at risk. As indicated elsewhere in this book such estimates must be extended beyond geographical pockets of homosexual and bisexual males in major urban areas. Furthermore, any hybrid modeling procedures developed for future studies need to include some of the mathematical modeling procedures in geographical diffusion studies by Cliff et al. (1981), Cliff, Haggett, & Ord (1986), and Pyle (1986).

As with attempts to describe other aspects of the geography of HIV infection and AIDS, the unavailability of adequate and accurate data precludes development of definitive population and geographic models (Gould, 1989). Yet, despite problems related to data availability and accuracy, there is no doubt about the substantial geographic differences in the distribution of populations infected with HIV and manifesting AIDS (Mann et al., 1986; Wendt, Sadowski, Markowitz, & Saravolatz, 1987; Piot et al., 1988). The variable geographic distribution of infection and related opportunistic diseases and conditions points to the importance of spatially variable behavioral, social, and environmental factors, and also suggests the utility of considering AIDS and the geography of AIDS within a socioecological model of disease (Armstrong, Gold, et al., 1985; Biggar, 1986; Calabrese, Proffitt, et al., 1986).

A socioecological model accounts for disease through the interaction of three major sets of factors (1) those related to the host, such as biological experience and genetic susceptibility; (2) external environmental factors including the presence or absence of disease pathogens, or the mutation and changing virulence of a pathogen; and (3) personal behavior factors that may prevent or limit exposure to a pathogen or, alternatively, facilitate transmission of a pathogen (Dever, 1980). Regardless of scale, the dynamic interrelationships between these sets of factors not only appear to be instrumental in understanding the geographic patterns of the AIDS pandemic, but also provide the basis for development of models that attempt to replicate and predict the diffusion of HIV infection and associated health problems.

We have seen, for example, how certain culturally sanctioned individual behaviors in the more remote parts of central Africa may have been sufficient to contribute to cross-species transfer of the HIV precursor from nonhuman to human primates. Without the presence of the pathogen in the environment, however, such behavior would be risk free with regard to development of the HIV.

Among intravenous drug abusers in some locations in the United States and in other countries, needle-sharing behavior is apparently an important part of the drug culture camaraderie and is linked strongly to diffusion of the HIV in some of the larger cities. Apparently, needle sharing is not as prevalent among drug users throughout much of Europe, resulting in a differing pattern of infection and associated diseases. However, even here geographic variations in drug-related behavior may lead to different infection routes. In Scotland, for example, an estimated 51% of injecting drug users in Edinburgh are reported to be infected with HIV, compared with less than 5% in Glasgow. Again, this has been attributed to differing patterns of needle and syringe sharing in the two drug-using communities (Browning, 1987).

Regardless of cultural background, sexual preference, or geographic location, it appears that sexual promiscuity greatly enhances the risk of HIV infection through increased exposure. The presence of particular genital diseases—host condition— renders some individuals more susceptible to infection.

The description of general patterns and the development of models that accurately replicate and predict the geographic pattern of infection are instrumental in forecasting the geographic dimensions and potential impact of the pandemic, in determining meaningful points of intervention to prevent continued diffusion, and in planning for the most efficient and appropriate distribution of treatment facilities (Brookmeyer & Gail, 1986; Anderson et al., 1987; Wood, 1988).

GLOBAL PATTERNS

Though fewer than 10 years have elapsed since identification of the HIV, on a global scale, three broad but apparently distinct patterns of infection have been distinguished, and a fourth recently suggested (Von Reyn & Mann, 1987; Mann, 1988; Torrey, Way & Rowe, 1988; Piot et al., 1988; Beach, Mantero-Atieuza, Fordyce-Baum, Prineas, et al., 1989).

Pattern I

The first pattern is found largely in sub-Saharan central, eastern, and southern Africa and increasingly in Latin America, especially in Haiti, and some areas in South America. In the majority of these locations HIV infection likely began to spread extensively during the mid- to late 1970s. Heterosexuals are the main population groups affected, and the primary mode of transmission is through heterosexual sexual activity. The male-to-female ratios of infection are approximately 1:1. In this pattern the spread by intravenous drug use is relatively rare, but the HIV may be spread by the repeated use of needles without sterilization and, less commonly, through the repeated use of other skin piercing instruments for medical or ritual purposes. Infection through the use of tainted blood products is also a predominant mode of infection. Because of the high percentage of women infected, perinatal transmission is a major problem in most of these areas.

Pattern II

The second pattern occurs throughout most of North America, Western Europe, Australia, New Zealand, and urban areas of South America such as Rio de Janeiro and Sao Paulo, Brazil. It is characterized by the likely origin and extensive spread of the HIV from the mid-1970s to early 1980s. In the initial wave of the epidemic most cases occur among homosexual males and homosexual intercourse is the primary mode of infection. Quite rapidly and almost simultaneously an epidemic infection wave courses through the intravenous-drug-user communities. Rates of heterosexual transmission, representing a small percentage of cases, are initially low but increase over time. Heterosexual transmission most often occurs from infected males, predominately intravenous drug users or bisexuals. The male-to-female ratio of infection is usually greater than 10:1. Blood products for transfusion are screened and essentially safe. Perinatal transmission, from mother to child, is uncommon because of the relatively few women thus far infected.

Pattern III

The third pattern is presently reflected in Eastern Europe, the Middle East, North Africa, most countries of Asia and Oceania, and some rural areas of South America. The HIV appears to have been introduced relatively recently, during the early to mid-1980s. These countries currently account for a small percentage of AIDS cases, less than 1% of the reported world total. Early infection or large concentrations of people developing AIDS are generally associated with transfusions of blood products from other areas, especially Pattern II countries. Both heterosexual and homosexual transmission has been documented. In most instances the source of early infection appears to derive from sexual contact with populations from Pattern I and II countries. Prostitutes are among the highest risk groups.

It should be noted that these patterns are broad generalizations and different patterns may coexist within a single country, or even within large individual metropolitan areas. Additionally, the patterns may be expected to change as the infection spreads through the populations of these countries (Mann, 1988). The simultaneous coexistence of one or more patterns perhaps underlies the suggestion of a possible fourth pattern of infection.

Pattern IV

Observation of the spread of HIV infection in Brazil and Honduras suggests a more important role of bisexual individuals' involvement in spreading the infection from homosexual males to the larger heterosexual population (Beach et al., 1989; Cortes, Detels, Aboulafia, et al., 1989; Hopesdales & Mahabir, 1988). In Honduras, for example, only 14% of AIDS cases describe themselves as homosexual and an additional 13% describe themselves as bisexual. Sixty-six percent of the cases have occurred among self-described heterosexuals and over 30% in women, a large majority of whom are not prostitutes. It is suggested that the majority of women were infected via bisexual males.

BROAD CONCEPTUAL MODELS OF GEOGRAPHIC DIFFUSION

Based primarily on observations of the African experience and infection patterns in the United States and Europe, Wood

(1988) offers three useful but rather broad geographic models of the HIV diffusion pattern.

The first discussed here is termed the "North Diffusion Pattern" developed to reflect the diffusion experience of Western or developed nations. Without speculating about the initial "seeding" or origin of the infection, this model focuses on an essentially closed national or regional urban system and the diffusion of the infection out of large urban areas where groups at high risk to HIV infection—including male homosexuals, intravenous drug abusers, and prostitutes—are more prevalent. Particularly important to this model is the mobility of sexually active homosexual males. The primary route of infection moves as homosexuals move between larger urban areas and smaller cities or towns. The hierarchical nature of the diffusion pattern from larger cities to smaller cities and towns is emphasized.

It may be useful to revise this model in an attempt to account for some additional spatiotemporal properties of human behavior associated with the transmission of HIV. For example, it may be profitable to examine differences in the spatiotemporal behavior of homosexuals/bisexuals, prostitutes, and intravenous drug users.

Homosexual/Bisexuals

Many larger cities not only have larger absolute numbers of homosexuals, but substantial proportions reside in rather well-defined and identified subcommunities. While there is some indication of behavioral change among homosexual males, the promiscuous high risk lifestyle of many has led to the intense circulation of the HIV infection in these geographically distinct communities (see Figure 8.1). This type of geographic clustering of sexual contact has been demonstrated for other sexually transmitted diseases such as gonorrhea (Rothenberg, 1983), syphilis, and *Chlamydia trachomatis* infections (Alvarez-Dardet, Marguez & Perea, 1985).

The attraction of the facilities and character of these communities goes beyond the immediate area, however. Assuming a distance–decay effect operating on travel, we might expect

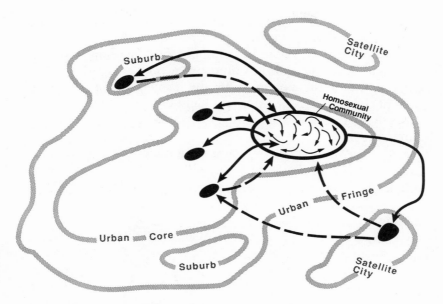

FIGURE 8.1. Postulated spatial model 1: HIV infection among male homosexuals and bisexuals.

homosexual and bisexual males from other parts of the city to visit the communities quite regularly. These frequent visits coupled with an intense level of risky behavior suggest the likelihood of a rather strong secondary path of infection among other homosexuals/bisexuals residing in the urban area. In the same manner a tertiary route of infection involving homosexuals or bisexuals in the more distant suburbs and smaller cities and towns can be hypothesized.

In addition to these major infection routes among homosexual and bisexual males in major metropolitan regions, an interregional and even national pattern must be considered. This would include homosexual and bisexual visitors from other large cities, smaller cities, and towns in the geographic interstices. From a temporal perspective, *ceteris paribus*, the model posits the initial and major infections in the largest cities, gradually diffusing downward to smaller cities and, eventually, to the smallest cities and towns. In fact, as we have seen, the first wave of the AIDS epidemic in the United States closely reflects this geographic pattern of involvement. For example, in a study of sexual contacts among homosexual men with AIDS at an early point in the epidemic in the United States, 40 AIDS patients in 10 different cities and eight different states were ultimately linked by sexual contact (Auerback, Darrow, Jaffe, & Curran, 1984).

Moreover, these patients were ultimately linked to a single index individual, Patient Zero, who developed lymphadenopathy in December 1979 and Kaposi's sarcoma in early 1980. Patient Zero's occupation as an international airline steward took him frequently from Europe to New York and Los Angeles. He estimated that he had approximately 250 different male sexual partners each year from 1979 through 1981. Of the 72 partners he was able to name, eight were AIDS patients: four from southern California and four from New York City. Among the 15 AIDS patients interviewed in the study, the average number of different sexual partners over the previous five-year period was 610.

Prostitutes

A similar but less intense geographic route of infection is associated with the location and use of prostitutes (Figure 8.2). To date, in most Western or developed countries prostitutes represent a relatively small but increasingly important source of HIV infection. In most cities prostitution is concentrated in certain areas. Again, the residential distribution of the patrons of prostitutes would be expected to reflect a distance–decay function. With increasing residential distance of patrons from the district, therefore, a decrease in the frequency of visit can be expected. Any model incorporating this decay function, therefore, would reflect a spread of infection similar to but much less intense than that posited for homosexual and bisexual males visiting predominantly homosexual residential and social sections of cities.

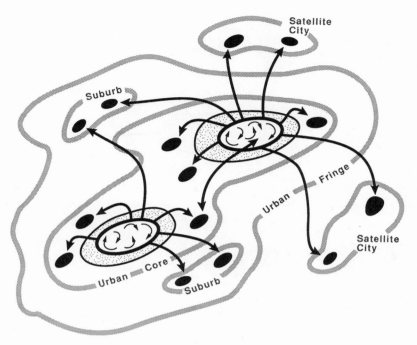

FIGURE 8.2. Postulated spatial model 2: HIV infection spread by prostitutes.

Intravenous Drug Users

In contrast to the broad geographic and temporal infection patterns of the previous two groups, we would expect a much more geographically restricted and localized pattern involving intravenous drug abusers (see Figure 8.3). Here it is important to distinguish between behavior patterns among various types of drug users. Again, in larger cities there are particular areas, generally located in lower socioeconomic and minority residential sections of cities, that are well known as distribution "centers" for various drugs. Control of these centers and their "service areas" frequently lead to battles for territorial rights. We also know that middle-class residents of the city and suburbs

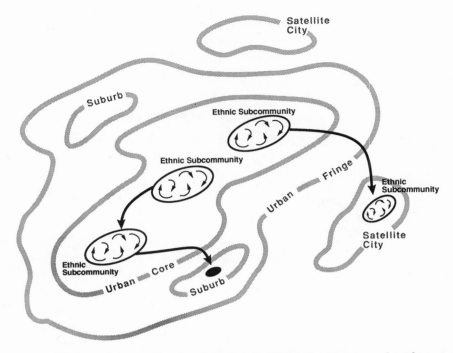

FIGURE 8.3. Postulated spatial model 3: HIV infection spread within ethnic subcommunities with IV drug abuse.

sometimes travel to these areas for their supply, but most often the drug they choose is not associated with intravenous injection. This practice appears limited primarily to lower socioeconomic and minority groups in the central cities. The sharing and repeated use of hypodermic syringes is often part of the drug culture among these groups. The nature of this drug use/addiction imposes a need for frequent use and associated resupply. This then causes intravenous drug abusers to be geographically concentrated in or very near supply centers and to limit their activity spaces and daily travel patterns. We would expect, therefore, a very severe distance–decay pattern of infection resulting from intravenous drug abuse centered on the drug supply centers.

To illustrate, a recent study of the distribution of AIDS cases by census tract in metropolitan Los Angeles County revealed several geographic clusters or extremely high numbers of cases (Greenwald, Terukina, & Weintraub, 1989). Two clusters were located near downtown Los Angeles and a third on the coast in Long Beach. Similarly a recent study of the Bronx, a borough of New York City, found over 60% of AIDS cases among intravenous drug users (Drucker & Vermund, 1989). Moreover, two-thirds of intravenous-drug-use-related AIDS deaths in the Bronx occurred in three South Bronx health districts. In so far as the intravenous drug users are sexually active, it is expected that the heterosexual spread of the infection will also be concentrated in this area.

Thus, we might expect in large central cities several endemic HIV foci associated with heavy concentrations of intravenous drug abuse and related sexual activity. These areas can be considered "ecological risk areas" which, depending on length of time spent in them as well as behavior while in these areas, represents "high impact" areas with regard to jeopardizing an individual's health (Medvedkov, 1969; Shannon & Spurlock, 1976). Travel patterns that bring individuals to these areas or, alternatively, keep individuals in these areas for the purpose of intravenous drug use and/or prostitution are thereby instrumental in exposing individuals to an increased risk of HIV infection. Just as importantly, therefore, these foci or risk areas may also serve as important points of origin for the heterosexual transmission of the HIV to other points within the community. As one epidemiologist has suggested, "it's not just what you do, it's where you do it."

In so far as there is a lack of intravenous drug use among middle- and upper-class residents of the city and surrounding suburbs and towns, we would expect that only relatively rarely would middle-class residents of the city and suburbs travel to intravenous drug supply centers in the central city *and* participate directly by sharing needles in the drug subculture there. However, the use of prostitutes in certain overlapping drug and prostitution risk areas by travellers from other areas of the community is well documented.

Given these considerations, therefore, it may be most useful if

individual models attempting to replicate and predict the spatial
and temporal diffusion of HIV infection incorporate information and informed assumptions about geographic characteristics and behavior of the groups. Without such consideration,
important elements of the diffusion process may be masked.
Subsequently, the individual models may be linked together to
provide a more integrated spatiotemporal perspective of the
locationally and behaviorally specific situations.

GEOGRAPHIC MODELING OF HIV INFECTION
IN DEVELOPING NATIONS

The second broad geographic model is the "South Diffusion
Pattern," based largely on the African experience. The primary
populations involved are sexually active and mobile heterosexuals, prostitutes, soldiers, and truck drivers. Although the role of
soldiers is somewhat ill-defined, research strongly supports the
important role of truck drivers and prostitutes. Emphasized
here is the movement between large urban areas and interven-
ing market towns situated along major connecting highways. A
secondary flow of infection is carried by migrants moving from
both the major urban areas and market towns to the more rural
villages. Again, there is no indication as to the source of the infection and the model represents the circulation and routes of
infection in a mature stage of the epidemic.

As with any attempt at modeling, the larger the scale considered, generally, the less applicable the resulting model to any
specific situation. Other modifications of the models might be
useful if they attempt to incorporate other observations of the
infection process. Alternatively, perhaps versions of the models
might be developed to deal more specifically with regional patterns. The "South" model, when applied to the sub-Saharan African experience, for example, might deal more specifically with
the possible transmission of infection from the areas surrounding the villages to the villages themselves and, subsequently,
from the villages to the larger towns and cities. From published
reports and as discussed in Chapter 4, this appears to reflect the

experience of Uganda, for instance. Additionally, this macro-scale model might also be adjusted to reflect another ack-nowledged important source of HIV infection in under-developed and developing countries, namely, contaminated blood products that might be imported internationally or de-rived from local populations and distributed locally and region-ally.

A COMBINED GEOGRAPHIC MODEL

A third model combines the North and South models to reflect migrations and, hence, diffusion of the infection across inter-national boundaries. The assumption in this model is that inter-national diffusion of infection occurs between large urban areas with strong international linkages. Recent evidence de-scribing a cluster of HIV infection among heterosexuals reflects the importance of this type of spatial diffusion route (Clumeck et al., 1989).

A cluster of 19 women (12 Europeans and 7 Africans) living in Antwerp and Brussels were identified as having had sexual con-tact with the same man, an HIV-infected civil engineer from a central African country. It is believed that the engineer, who traveled frequently between Brussels and central Africa, is re-sponsible for infecting up to seven of the European and four of the African women with whom he had sexual contact. Of the 11 infected women, seven had had sex with the index patient dur-ing the three years before his death. He reportedly averaged about 20 female sexual partners per year. Similar so-called "high disseminator" heterosexual males have been described in California and Sweden (Padian, Marquis, Francis, et al., 1987; Franzen, Alber, Biberfeld, Lidin-Janson, & Lowhagen, 1988).

OVERALL MODELING OF THE GEOGRAPHY OF THE HIV AND AIDS

The extant complex spatial mobility and interaction patterns probably guarantee the futility of searching for a single model to replicate the past geographic development or future path of HIV diffusion. Thus, the approach by Wood (1988) in developing models associated with major identified broad geographic and social patterns of infection is well founded. These models provide the basis for understanding the geographic diffusion patterns at the broadest levels. However, it is also useful to examine these broad patterns more carefully and to account for major geographic and spatial dimensions of significant actors and factors known to contribute to the diffusion of the infection. For example, we have suggested that there may be major distinctions between the geographic location and spatial interaction patterns or activity spaces of homosexuals/bisexuals, prostitutes, and intravenous drug abusers. Because of these differences, more comprehensive and geographically "sensitive" models of HIV-infection paths might be developed. To provide the most accurate representation, ultimately these models must be linked to models replicating the diffusion of HIV infection among subgroups in the general population with known behaviorally related risk probabilities of infection. Furthermore, the mathematical "back calculation" method can be used in such multiple modeling procedures.

To be most effective, attempts to model the development of HIV infection must account for the major infection routes within an area, both social and geographic. In addition, the models must take into account the stage of the epidemic and the geographic scale of investigation. Of paramount importance is the concentrated and expeditious development of these models if their forecasts are to have any significance to the amelioration of a very threatening situation. In this regard, the work of Drucker and Vermund (1989) is particularly notable. Without a major modeling effort and the effective integration of the results into related educational and health policy, program development, and implementation, there is a strong possibility the modeling will remain little more than an academic exercise.

References

Adelman, M. (1988). Social support and AIDS. *AIDS and Public Policy Journal, 4*, 31–39.

Adler, M.W. (1987). Development of the epidemic. *The British Medical Journal, 294*, 1083–1085.

Airhihenbuwa, C.O. (1989). Perspectives on AIDS in Africa: Strategies for prevention and control. *AIDS Education and Prevention, 1*, 57–69.

Alvarez-Dardet, C., Marquez, S., & Perea, E.J. (1985). Urban clusters of sexually transmitted diseases in the city of Seville, Spain. *Sexually Transmitted Diseases*, July–September, 166–168.

Anderson, R.M., May, R.M., & McLean, A.R. (1988). Possible demographic consequences of AIDS in developing countries. *Nature, 332*, 228–233.

Anderson, R.M., Medley, G.F., Blythe, S., & Johnson, A. (1987). Is it possible to predict the minimum size of the acquired immunodeficiency syndrome (AIDS) in the United Kingdom? *The Lancet, 1*, 1073–1075.

Andre, L. (1987). Le S.I.D.A. À-t-il Déjà existé? *Médecine Tropicale, 47*, 229–230.

Andrulis, D.P., Weslowski, V.B., & Gage, L.S. (1989). The 1987 US hospital AIDS survey. *Journal of the American Medical Association, 262*, 784–794.

Andrulis, D.P., et al. (1987). The provision of financing of medical care for AIDS patients in the U.S. public and private teaching hospitals. *Journal of the American Medical Association, 258,* 1343–1346.

Anonymous (1986). Competition and incompetence hampers AIDS research. *New Scientist, 112,* 7.

Anonymous (1987a). Austria acts against AIDS. *New Scientist, 113,* 27.

Anonymous (1987b). Immigration guidelines set to change. *New Scientist, 113,* 26.

Anonymous (1988a). The USSR and the origin of the AIDS virus. *AIDS: Global showdown: Mankind's total victory or total defeat. Part VII* (pp. 119–126). Washington, DC: Executive Intelligence Review News Service.

Anonymous (1988b). Royal Statistical Society Meeting on AIDS. *Journal of the Royal Statistical Society (Ser. A), A151,* Pt.1, 1–131.

Anonymous (1989a). AIDS: Prevention, policies, and prostitutes. *The Lancet, i,* 1111–1113.

Anonymous (1989b). Interpretation and use of the Western blot assay for serodiagnosis of human immunodeficiency virus Type 1 infections. *Journal of the American Medical Association, 262,* 3395–3397.

Anonymous (1989c). AIDS update. *British Medical Journal, 298,* 1057.

Anonymous (1990). AIDS in Thailand. *British Medical Journal, 300,* 415–416.

Armstrong, D., Gold, J., et al. (1985). Treatment of infections in patients with the acquired immunodeficiency syndrome. *Annals of Internal Medicine, 103* 738–743.

Aubry, P. (1989). Le SIDA dans les Caraibes et en Oceanie. *Médecine Tropicale, 49,* 9–10.

Auerbach, D.M., Darrow, W.W., Jaffe, H.W., & Curran, J.W. (1984). Cluster of cases of the acquired immune deficiency syndrome: Patients linked by sexual contact. *The American Journal of Medicine, 76,* 487–492.

Auti, F. et al. (1985). IgM and IgG antibodies to human t cell lymphotropic retrovirus (HTLV-III) in lymphadenopathy syndrome and subject at risk in Italy. *British Medical Journal, 291,* 165–166.

Baltimore, D., & Feinberg, M.B. (1989). HIV revealed: Toward a natural history of the infection. *New England Journal of Medicine, 321,* 1673–1675.

Barker, C., & Turshen, M. (1986). AIDS in Africa. *Review of African Political Economy, 36,* 51–54.

Barry, M., Mellors, J., & Bia, R. (1984). Haiti and the AIDS connection. *Journal of Chronic Diseases, 37,* 592–595.

Barton, T. (1988). Sexually-related illness in Eastern and Central Africa: A selected bibliography. In N. Miller & R. Rockwell (Eds.), *AIDS in Africa: The Social and Policy Impact.* (pp. 269–291). Lewiston, NY: E. Mellen Press.

Beach, R.S., Mantero-Atienza, E., Fordyce-Baum, M.K., et al. (1989). HIV infection in Brazil. *The New England Journal of Medicine, 321,* 830.

Beardsley, T. (1986). AIDS research queried: Skulduggery at the lab bench. *Nature, 324,* 506.

Benjamin, A. (1988). Long-term care and AIDS: Perspectives from experience with the elderly. *Milbank Memorial Quarterly, 66,* 415–443.

Bennett, F.J. (1962). The social determinants of gonorrhoea in an East African Town. *East African Medical Journal, 39,* 332–342.

Beral, V., Peterman, T., Berkelman, R., & Jaffe, H. (1990). Kaposi's sarcoma among persons with AIDS: A sexually transmitted infection? *The Lancet, 335,* 123–128.

Biggar, R.J. (1986). The clinical features of HIV infection in Africa. *British Medical Journal, 293,* 1453–1454.

Biggar, R.J. (1988). Overview: Africa, AIDS, and Epidemiology. In N. Miller & R. Rockwell (Eds.) *AIDS in Africa: The Social Policy and Impact* (pp. 1–29). Lewiston, NY: E. Mellen Press.

Biggar, R.J. (1987). AIDS in subSaharan Africa. *Cancer Detection and Prevention Supplement, 1:* 487–491.

Biggar, R.J., Melbye, M., et al. (1985). ELISA HTLV retrovirus antibody reactivity associated with malaria and immune complexes in healthy Africans. *The Lancet, 2,* 520–523.

Bohlen, C. (1990, February 8). Romania's AIDS babies: A legacy of neglect. *The New York Times,* pp. 1, 12.

Bongaarts, J., Reining, P., Way, P., & Conant, F. (1989). The relationship between male circumcision and HIV infection in African populations. *AIDS, 3,* 373–377.

Bouvet, E., et al. (1983). Sarcome de Kaposi et infectious opportunistes chez des sujets jeunes sans Antécédent susceptible d'eutraianer une immuno-dépression. *Le Presse Médicate, 12,* 2431–2434.

Brokensha, D. (1988). Overview: Social factors in the transmission and control of African AIDS. In N. Miller & R.C. Rockwell (Eds.), *AIDS in Africa: The Social and Policy Impact* (pp. 167–173). Lewiston, NY: E. Mellen Press.

Brookmeyer, R. & Gail, M.H. (1986). Minimum size of the acquired immunodeficiency syndrome (AIDS) epidemic in the United States. *The Lancet, ii,* 1320–1322.

Brookmeyer, R., & Gail, M.H. (1988). A method for obtaining short-term projections and lower bounds on the size of the AIDS epidemic. *Journal of the American Statistical Association, 38* 301–308.

Brown, L.A. (1981). *Innovation diffusion: A new perspective.* New York & London: Methuen.

Browning, M.J. (1987). AIDS in Scotland: An update. *Scottish Medical Journal, 32,* 4–15.

Bruckner, G., Brun-Vézinet, F., Rosenheim, M., et al. (1987). HIV-2 infection in two homosexual men in France. (letter) *The Lancet, i* 223.

Brunet, J.B., & Ancelle, R.A. (1985). The international occurrence of the acquired immunodeficiency syndrome. *Annals of Internal Medicine, 103,* 670–674.

Brunet, J.B., Boubet, E., & Leibowitch, J. (1983). Acquired immunodeficiency syndrome in France. *The Lancet, i,* 700–701.

Brun-Vézinet, F., Rouzioux, C., Moutagnier, L., et al. (1984). Prevalence of antibodies to lymphadenopathy-associated retrovirus in African patients with AIDS. *Science, 226,* 453–46.

Bucknall, A. (1986). Regional patterns of AIDS and HIV infection. *Journal of the Royal College of General Practitioners, 36,* 491–492.

Bureau of Intelligence and Research. (1990). AIDS in the USSR. *Geographic Notes,* No. 11, 6–7.

Bygbjerg, I.C. (1983). AIDS in a Danish surgeon (Zaire, 1976). *The Lancet, i,* 925.

Cahill, K. (1984). *The AIDS epidemic.* London: Hutchinson.

Calabrese, L.H., Proffitt, M.R., et al. (1986). Epidemiologic and laboratory evaluation of homosexual males from an area of low incidence for acquired immunodeficiency syndrome (AIDS). *Cleveland Clinic Quarterly, 53,* 267–275.

Castro, K.D., Lieb, S., et al. (1988). Transmission of HIV in Belle Glade, Florida: Lessons for other communities in the United States. *Science, 239* 193–197.

Centers for Disease Control. (1981a). Pneumocystis pneumonia—Los Angeles. *Morbidity and Mortality Weekly Report, 30,* 250–252.

Centers for Disease Control. (1981b). Kaposi's sarcoma and pneumocystis pneumonia among homosexual men—New York City and California. *Morbidity and Mortality Weekly Report, 31,* 465–467.

Centers for Disease Control. (1987a). Revision of the CDC Surveillance case definition for acquired immunodeficiency syndrome. *Morbidity and Mortality Weekly Report, 36, Suppl. No. S-1,* August 14.

Centers for Disease Control. (1987b). Human immunodeficiency virus (HIV) infection codes: Official authorized addendum ICD-9-CM effective January 1, 1988. *Morbidity and Mortality Weekly Report, 36,* 1–24.

Centers for Disease Control. (1987c). Antibody to human immunodeficiency virus in female prostitutes. *Morbidity and Mortality Weekly Report, 36,* 157–161.

Centers for Disease Control. (1987d). Update: Human immunodeficiency virus infection in health-care workers exposed to blood or infected patients. *Morbidity and Mortality Weekly Report, 36,* 285-289.

Centers for Disease Control. (1987e). Recommendation for prevention of HIV transmission in health care settings. *Morbidity and Mortality Weekly Report, 36, Suppl. 2S.*

Centers for Disease Control. (1987f). Human immunodeficiency virus infection in the United States: A review of current knowledge. *Morbidity and Mortality Weekly Report, 36, Suppl. No. S-6,* December 18.

Centers for Disease Control. (1988a). AIDS due to HIV-2 infection— New Jersey. *Morbidity and Mortality Weekly Report, 37,* 33-35.

Centers for Disease Control. (1988b). Transmission of HIV through bone transplantation: Case report and public health recommendations. *Morbidity and Mortality Weekly Report, 37,* 597-599.

Centers for Disease Control. (1989a). AIDS and human immunodeficiency virus infection in the United States: 1988 update. *Morbidity and Mortality Weekly Report, 38, Suppl. No. S-4.*

Centers for Disease Control. (1989b). Update: Heterosexual transmission of acquired immunodeficiency syndrome and human immunodeficiency virus infection: United States. *Morbidity and Mortality Weekly Report, 38,* 423-434.

Centers for Disease Control. (1990a). HIV/AIDS surveillance report. January.

Centers for Disease Control. (1990b). Estimates of HIV Prevalence and Projected AIDS Cases. *Morbidity and Mortality Weekly Report, 39,* 110-119.

Chiodi, F., Biberfeld, G., Parks, E., et al. (1989). Screening of African sera stored for more than 17 years for HIV antibodies by site-directed serology. *European Journal of Epidemiology, 5,* (1), 42-46.

Clavel, F., Mansinho, K., & Chamanet, S. (1987). Human immunodeficiency virus type 2 infection associated with AIDS in West Africa. *The New England Journal of Medicine, 316,* 1180-1185.

Cliff, A.D., et al. (1981). *Spatial diffusion: An historical geography of epidemics in an island community.* New York: Cambridge University.

Cliff, A.D., Haggett, P., & Ord, J.K. (1986). *Spatial aspects of influenza epidemics.* London: Dion.

Clumeck, N. (1984). Acquired Immune Deficiency Syndrome in Belgium. *European Journal of Clinical Microbiology, 3,* 59-60.

Clumeck, N., & De Wit, S. (1988). AIDS in Africa. In M. Sande & P. Volberding (Eds.) *The medical management of AIDS* (pp. 331-349). Philadelphia: Saunders.

Clumeck, N., Sonnet, J., Taelman, H., et al. (1984). Acquired immuno-deficiency syndrome in African patients. *The New England Journal of Medicine*, 310, 492–497.

Clumeck, N., Taelman, H., Hermans, et al. (1989). A cluster of HIV infection among heterosexual people without apparent risk factors. *The New England Journal of Medicine*, 321, 1460–1462.

Clumeck, N., Van De Perre, P., Carael, M., et al. (1985). Heterosexual promiscuity among African patients with AIDS. *The New England Journal of Medicine*, 312, 182.

Conway, G.A., Colley-Niemeyer, B., Pursley, C., et al. (1989). Under-reporting of AIDS cases in South Carolina, 1986 and 1987. *Journal of the American Medical Association*, 262, 2859–2863.

Coombs, R.W., Collier, A.C., Allain, J.P., et al. (1989). Plasma viremia in human immunodeficiency virus infection. *The New England Journal of Medicine*, 321 1626–1631.

Cordes, C. (1990). End to special status of AIDS studies at NIH being sought. *The Chronicle of Higher Education*, 36, A1, A28–9.

Cortes, E., Detels, R., Aboulafia, D., et al. (1989). HIV-1, HIV-2, and HTLV-I infection in high-risk groups in Brazil. *The New England Journal of Medicine*, 302, 1005–7.

Crewdson, J. (1990, March 18). Inquiry hid facts on AIDS research. *Chicago Tribune*, 143, 77: pp. 1, 12–13.

Crosette, B. (1990, February 13). The 42 poorest nations plan a campaign for help. *The New York Times*, p. A3.

Curran, J.W. (1989). Trends in the epidemiology of HIV-related illnesses: Implications of recent research. In *New Perspectives on HIV-Related Illnesses: Progress in Health Services Research* (pp. 7–12). National Center for Health Services Research and Health Care Technology Assessment, Public Health Service, U.S. Department of Health and Human Services.

Dawson, M.H. (1988). AIDS in Africa: Historical roots. In N. Miller & R. Rockwell (Eds.), *AIDS in Africa: The social and policy impact* (pp. 57–69). Lewiston, NY: E. Mellen Press.

Dechau, C.P. (1987). Compulsory AIDS testing in Bavaria. *Nature*, 326, 4.

Denis, F., Gershy-Damet, G., Lhuillier, M., et al. (1987). Prevalence of human T-lymphotropic retroviruses type III (HIV) and Type IV in Ivory Coast. *The Lancet, i*, 408–11.

DePerre, P., Lepage, P., et al. (1984). Acquired immunodeficiency syndrome in Rwanda., *The Lancet, ii*, 62–69.

Desmyter, J., Surmont, I., Goubau, P., & Vandepitte, J. (1986). Origins of AIDS. *British Medical Journal*, 293: 1306.

Dever, G.E.A. (1980). *Community Health Analysis*. Germantown, MD: Aspen Systems Corporation.

Dickson, D. (1987a). Africa begins to face up to AIDS. *Science, 238,* 605–607.

Dickson, D. (1987b). France, Britain boosts AIDS funds. *New Scientist, 235,* 1136.

Doolittle, R.F. (1989). The simian-human connection. *Nature, 339,* 338–339.

Drucker, E., & Vermund, S.H. (1989). Estimating population prevalence of human immunodeficiency virus infection in urban area with high rates of intravenous drug use: A model of the Bronx in 1988. *American Journal of Epidemiology, 130,* 133–142.

Duesberg, P.H. (1987). Retroviruses as carcinogens and pathogens: Expectations and reality. *Cancer Research, 47,* 1199–1220.

Duesberg, P.H. (1989). Human immunodeficiency virus and acquired immunodeficiency syndrome: Correlation but not causation. *Proceedings of the National Academy of Science of the United States of America, 86,* 755–764.

Dufoort, G., Courouce, A.M., Ancelle-Park, R., & Bletry, O. (1988). No clinical signs 14 years after HIV-2 transmission via blood transfusion. *The Lancet, ii,* 510.

Dutt, A. et al. (1988). Geographical patterns of Aids in the United States. *The Geographical Review, 77,* 456–471.

Ellrodt, A., Le Bras, P., et al. (1984). Isolation of human T-lymphotropic retrovirus (LAV) from Zairian married couple, one with AIDS, one with prodomes. *The Lancet, ii,* 1383–1385.

Ensolli, B., Barillari, G., Salahuddin, S., Gallo, R., & Wong-Staal, F. (1990). Tat protein of HIV-1 stimulates growth of cells derived from Kaposi's sarcoma lesions of AIDS patients. *Nature, 345:* 84–86.

Essex, M. (1989). Origins of AIDS. In V.T. DeVita, Jr., S. Hellman, & S.A. Rosenberg (Eds.), *AIDS: Etiology, diagnosis, treatment, and prevention* (pp. 3–10). Philadelphia: Lippincott.

Essex, M., & Kanki, P. (1988). The origins of the AIDS virus. *Scientific American, 259,* 64–71.

Evans, B., McClean, K., Dawson, S., et al. (1989). Trends in sexual behavior and risk factors for HIV infection among homosexual men, 1984–7. *British Medical Journal, 298,* 215–221.

Farthing, C., Brown, S., & Staughton, R. (1988). *Color atlas of AIDS and HIV diseases* (2nd ed.). Chicago: Yearbook Medical Publishers.

Fauci, A.S. (1988). The human immunodeficiency virus: Infectivity and mechanisms of pathogenesis. *Science, 239,* 617–622.

Fink, A.J. (1986). A possible explanation for heterosexual male infection with AIDS. *The New England Journal of Medicine, 315,* 1167.

Fourcroy, J.L. (1983). L'éternal couteau: Review of female circumcision. *Urology, XXII,* 458–461.

Fox, E., Abatte, E.A., Salah, S., Constantine, N.T., Rodier, G., & Woody, J.N. (1989). Incidence of HIV infection in Djibouti in 1988. *AIDS, 3,* 244–245.

Franzen, C., Albert, J., Biberfeld, G., et al. (1988). Natural history of heterosexual HIV infection in a Swedish cohort. In *Abstracts of the Fourth International Conference on AIDS.* Stockholm: Swedish Ministry of Health and Social Affairs.

Gail, M.H., & Brookmeyer, R. (1988). Methods for projecting course of acquired immunodeficiency syndrome epidemic. *Journal of the National Cancer Institute, 80,* 900–911.

Gallo, R.C., Sarngadharan, M.G., et al. (1987). Human retroviruses with emphasis on HTLV-III/LAV: Now and future perspectives. In J.C. Gluckman and E. Vilmere (Eds.), *Acquired Immunodeficiency Syndrome* (pp. 127–130). Amsterdam: Elsevier.

Gallo, R.C., & Montagnier, L. (1988). AIDS in 1988. *Scientific American, 259,* 41–48.

Gardner, J.I., Jr., et al. (1989). Spatial diffusion of the human immunodeficiency virus infection epidemic in the United States 1985–87. *Annals of the Association of American Geographers, 79,* 25–43.

Gelderbloom, H.R., Hausman, E., et al. (1987). Fine structure of the human immunodeficiency virus (HIV) and immunolocalization of structural proteins. *Virology, 156,* 171–176.

Genin, C., Coulanges, P., & Rasamindrakotroka, A. (1988), *Bilan de l'action* menée par l'institut de Madagascar dans la recherche des sséropositivités au virus de l'immunodéficience humaine (VIH) dans ce pays. *Médecine Tropicale, 48,* 397–399.

Glenny, M. (1987). AIDS in the eastern bloc: On the enlightened attitude in Budapest. *New Scientist,* 113–61.

Golub, A., & Gorr, W. (1990). *The effect of spatial aggregation on reported growth of AIDS cases.* Carnegie Mellon University School of Urban and Public Affairs, Working Papers 90–95.

Good, C. (1988). Traditional healers and AIDS management. In N. Miller & R.C. Rockwell (Eds.), *AIDS in Africa: The social and policy impact* (pp. 97–113. Lewiston, NY: E. Mellen Press.

Gonzales, J.J., & Koch, M.G. (1987). On the role of transients (biasing transitional effects) for the prognostic analysis of the AIDS epidemic. *American Journal of Epidemiology, 126,* 985–1005.

Gonzalez, J.P., Georges-Courbot, M., Martin, P.M., et al. (1987). True HIV-1 infection in a pygmy. *The Lancet, i,* 1498.

Gould, P. (1989). Geographic dimensions of the AIDS epidemic. *The Professional Geographer, 41,* 71–78.

Gould P., Gorr, W., & Casetti, E. (1988). *Understanding and predicting the AIDS epidemic in geographic space,* (Mimeo). The Penn State–Carnegie Mellon–Ohio State Consortium.

Greenwald, A., Terukina, D., & Weintraub, D. (1989). AIDS in L.A.

Occasional Publications in Geography (No. 4). Northridge, CA: Department of Geography, California State University at Northridge.

Gruson, L. (1988, October 20). AIDS spreading in Central America. *The New York Times*, p. 13.

Haseltine, W.A., 7 Wong-Staal, F. (1988). The molecular biology of the AIDS virus. *Scientific American, 259*, 52–62.

Haq, C. (1988). Management of AIDS patients: Case report from Uganda. In N. Miller & R.C. Rockwell (Eds.), *AIDS in Africa: The social and policy impact* (pp. 87–94). Lewiston, NY: E. Mellen Press.

Hayes, J.A., Jr., Marlink, R.G., & Harawi, S.J. (1989). HIV-related disease in Africa. In S.J. Harawi & C.J. O'Hara (Eds.) *Pathology and pathophysiology of AIDS and HIV-related diseases* (pp. 443–458). St. Louis: C.V. Mosby.

Herskovits, J. (1971). *Life in a Haitian valley.* New York: Doubleday/ Anchor.

Hilts, P. (1990a, January 1). Forecast of AIDS cases is cut by 10%. *The New York Times*, p. 14.

Hilts, P. (1990b, February 8). W.H.O. emergency team is sent to Romania to assess AIDS cases. *The New York Times*, p. A12.

Hirch, V.M., Olmstead, R.A., Murphey-Corb, M., Purcell, R., & Johnson, P.R. (1989). An African primate lentivirus (SIVsm) closely related to HIV-2. *Nature, 339*, 389–392.

Ho, D.D., Moudgil, T., & Alam, M. (1989). Quantitation of human immunodeficiency virus type 1 in the blood of infected persons. *The New England Journal of Medicine, 321*, 1521–1525.

Homsy, J., Meyer, M., Tateno, M., et al. (1989). The Fc and not CD4 receptor mediates antibody enhancement of HIV infection in human cells. *Science, 244;* 1357–1360.

Hopesdales, C.J., & Mahabir, S. (1988). The epidemiology of AIDs in the Caribbean and action to date. In A.F. Fleming, M. Carballo, D.W. Fitzsimons, et al. (Eds.), *The Global Impact of AIDS* (pp. 27–33). New York: Alan R. Liss.

Horsburgh, C.R., Jr., & S.D. Homberg. (1988). The global distribution of human immunodeficiency virus type 2 (HIV-2) infection. *Transfusion, 28;* 192–195.

Hoyle, F., & Wickramasinghe, N.C. (1990). Sunspots and influenza. *Nature, 343,* 304.

Hrdy, D.B. (1987). Cultural practices contributing to the transmission of human immunodeficiency virus in Africa. *Reviews of Infectious Diseases, 9,* 1109–1119.

Ibrahim, Y. (1990, February 19). Culture and stigma slow AIDS reports in Mideast. *The New York Times*, p. A5.

Jaffe, H.W., & Lifson, A.R. (1988). Acquisition and transmission of

HIV. In M. Sande & P.A. Volberding (Eds.), *The Medical Management of AIDS* (pp. 19–27). Philadelphia: Saunders.

Johnson, L. (1989). Home care's challenge: Compassionate care for persons with AIDS. *Caring, 8,* 24–28.

Johnson, A., & Laga, M. (1990). Heterosexual transmission of HIV. In N. Alexander, H. Gabelnick, & J. Spieler (Eds.), *Heterosexual transmission of AIDS* (pp. 9–24). New York: Wiley-Liss.

Jougla, E., Hatton, F., Michel, E., & Letoullec, A. (1989). AIDS mortality in France (1983–1988). Unpublished manuscript presented at a workshop, "Modeling the Spread of HIV/AIDS and Its Demographic and Social Consequences," Budapest, Hungary, November 1989.

Kalyanaraman, V.S., Sarngadharan, M.G., & Gorodff, M. (1982). A new sub-type of human t-cell leukemia virus (HTLV-II) associated with a cell variant of hairy cell leukemia. *Science, 218,* 571–573.

Kanki, P., Alroy, J., & Essex, M. (1985). Isolation of T-lymphotropic retrovirus related to HTLV-III/LAV from wild caught African green monkeys. *Science, 230,* 951–954.

Kanki, P., M'Boup, S., Ricard, D., et al. (1987). Human T-lymphotropic virus type 4 and the human immunodeficiency virus in West Africa. *Science, 236,* 827–831.

Kantner, H., & Pankey, G. (1987). Evidence for a Euro-American origin of human immunodeficiency virus (HIV). *Journal of the National Medical Association, 79,* 1068–1072.

Kashamura, A. (1973). *Famille, sexualité et culture: Essai sur les moueurs sexuelles set les cultures des peuples des grands lacs Africains.* Paris: Payot.

Kawamura, M., Yamazaki, S., Ishikawa, K., et al. (1989). HIV-2 in West Africa in 1966. *The Lancet, i,* 385.

Kitchen, L.W. (1987). AIDS in Africa: Knowns and unknowns. *CSIS Africa Notes, 74,* 1–4.

Klatzman, D., & Gluckman, J.C. (1987). The pathophysiology of HIV infection: A complex pathway of host-virus interaction. In J.C. Gluckman & E. Vilmer (Eds.), *Acquired Immunodeficiency Syndrome,* (pp. 77–85). Amsterdam: Elsevier.

Knox, E.G. (1986). A transmission model for AIDS. *European Journal of Epidemiology June 5, 2,* 165–177.

Kolata, G. (1988). AIDS virus found to hide in cells, eluding detection by normal tests. *The New York Times,* p. 1, 15.

Konings, E., Anderson, R.M., Morley, D., O'Riordan, & Megan, M. (1989). Rates of sexual partner change among two pastoralist southern nilotic groups in East Africa. *AIDS, 3,* 245–247.

Konotey-Ahula, F.I.D. (1987). AIDS in Africa: Misinformation and disinformation. *The Lancet, 2,* 206–207.

Kreiss, J.K., Koech, D., Plummer, F.A., et al. (1986). AIDS virus infection in Nairobi prostitutes: Spread of the epidemic to East Africa. *The New England Journal of Medicine, 312,* 414–418.

Kuhls, T.L., Nishanian, P.G., Cherry, J.D., et al. (1988). Analysis of false positive HIV-1 serologic testing in Kenya. *Diagnostic Microbiology of Infectious Diseases, 9,* 179–186.

L'Age-Stehr, J. (1984). Acquired Immune Deficiency Syndrome in the Federal Republic of Germany. *European Journal of Clinical Microbiology, 3,* 61.

Latif, A.S., Katzenstein, D.A., Bassett, M.T., et al. (1989). Genital ulcers and transmission of HIV among couples in Zimbabwe. *AIDS, 3,* 519–523.

Leibowitch, J. (1985). *A strange virus of unknown origin.* (Translated from the French by R. Howard.). New York: Ballantine Books.

Lemaître, M., Guétard, D., Henin, Y., Montagnier, L., and Zerial, A. (1990). Protective activity of tetracycline analogs against the cyclopathic effect of the human immunodeficiency virus in CFM cells. *Research in Virology, 141:* 5–16.

Leonidas, J., & Hyppolite, N. (1983). Haiti and the acquired immunodeficiency syndrome. *Annals of Internal Medicine, 98,* 1021.

Letvin, N., Daniel, N., King, L., et al. (1988). An HIV-related virus from macaques. *AIDS, 2,* 71–74.

Levy, J.A. (1988). The human immunodeficiency virus and its pathogenesis. In M.A. Sande & P. Volberding (Eds.), *The medical management of AIDS* (pp. 5–17).Philadelphia: Saunders.

Levy, J.A. (1989). Human immunodeficiency viruses and the pathogenesis of AIDS. *Journal of the American Medical Association, 261,* 2997–3006.

Li, Y., Naidu, Y.M., Daniel, M. & Desrosiers, R.C. (1989). Extensive genetic variability of simian immunodeficiency virus from African green monkeys. *Journal of Virology, 63,* 1800–1802.

Lifson, A.R. (1988). Do alternate modes for transmission of human immunodeficiency virus exist? *Journal of the American Medical Association, 259,* 1353–1356.

Lightfoot-Klein, H. (1990). *Prisoners of ritual: An odyssey into female genital circumcision in Africa.* Binghamton, NY: Haworth Press.

Lyons, S., Schoub, B., & McGillivray, G. (1985). Lack of evidence of HTLV-III in Southern Africa. *The New England Journal of Medicine, 314,* 1257 1258.

Mabey, C.C., Tedder, R.S., et al. (1988). Human retroviral infections in the Gambia. *The British Medical Journal, 296,* 83–86.

Mann, J.M. (1987a). The global AIDS situation. *World Health Statistical Quarterly, 40,* 185–192.

Mann, J.M. (1988). The global picture of AIDS. *Journal of Acquired*

Immune Deficiency Syndrome, 1, 209–216.

Mann, J.M., Chin, J., et al. (1988). The international epidemiology of AIDS. *Scientific American, 259*, 82–89.

Mann, J.M. (1987b). The epidemiology of LAV/HTLV-III in Africa. In J.C. Gluckman & E. Vilmer (Eds.), *Acquired immunodeficiency syndrome* (pp. 131–157). Amsterdam: Elsevier.

Mann, J.M., Francis, H., Quinn, T., et al. (1986). Surveillance for AIDS in a Central African city: Kinshasa, Zaire. *The Journal of the American Medical Association, 255*, 3255–3259.

Mansell, P. (1988). AIDS: Home, ambulatory, and palliative care. *Journal of Palliative Care, 4*, 29–33.

May, R., & Anderson, R.M. (1987). Transmission dynamics of HIV infection. *Nature, 326*, 137–142.

McClure, M., & Schulz, T. (1989, May 13). Origin of HIV. *British Medical Journal, 298*, 1267–1268.

McEvoy, M. (1984). Acquired immune deficiency syndrome in the United Kingdom. *European Journal of Clinical Microbiology, 3*, 53–64.

McKusick, L., Conant, M., & Coates, T.J. (1985). The AIDS epidemic: A model for developing intervention strategies for reducing high-risk behavior in gay men. *Sexually Transmitted Diseases, 12*, 229–233.

McLean, S., & Graham, S.E. (Eds.). (1985). *Female circumcision, excision and infibulation: The facts and proposals for change* (Minority Rights Group Report No. 47). London: Murray House.

Medvedkov, U. (1969). Ecological risk areas. World Health Organization.

Metraux, A. (1958). *Le vaudou Haimtien*. Paris: Gallimard.

Miller, N., & Rockwell, R.C. (1988). Introduction. In N. Miller & R.C. Rockwell (Eds.), *AIDS in Africa: The social and policy impact* (pp. xxvii–xxxi). Lewiston, NY: E. Mellen Press.

Melbye, M. Bayleyi, A., Manuwelle, J.K., et al. (1986). Evidence for heterosexual transmission and clinical manifestations of human immunodeficiency virus infection and related conditions in Lusaka, Zambia. *The Lancet, 3*, 1113–1115.

Montagnier, L., & Alizon, M. (1987). The human immune deficiency virus (HIV): An update. In J.C. Gluckman & E. Vilmer (Eds.), *Acquired immunodeficiency syndrome* (pp. 13–21). Amsterdam: Elsevier.

Moore, A., & LeBaron, R. (1986). The case for a Haitian origin of AIDS epidemic. In D. Feldman & T. Johnson (Eds.) *The social dimensions of AIDS: Method and theory* (pp. 77–93). New York: Praeger.

Mortimer, P.P., et al. (1985). Prevalence to human t-lymphotropic virus type III by risk group and area, United Kingdom 1978–84. *British Medical Journal, 290,* 1176–1178.

Moss, A.R., & Bachetti, P. (1989). Natural history of HIV infection. *AIDS, 3,* 55–61.

Mulder, C. (1988). Human AIDS virus not from monkeys. *Nature, 333,* 396.

Murdock, G.P. (1967). Ethnographic atlas: Summary. *Ethnology, 6,* 107.

Nahmias, A.J., Weiss, J., Yao, X., et al. (1986). Evidence for human infection with an HTLVIII/LAV-like virus in Central Africa, 1959. *The Lancet, 1,* 1279–1280.

Najera, R., Herrera, M.I., & Andres, R. (1987). Human immunodeficiency virus and related retroviruses. *The Western Journal of Medicine, 147,* 702–708.

Neff, J. (1987). AIDS registration becoming a political issue in Germany. *Nature, 325,* 650.

Nemeth, A., Bygdeman, S., Sandstrom, E., & Biberfeld, G. (1986). Early case of acquired immunodeficiency syndrome in a child from Zaire. *Sexually Transmitted Diseases,* April–June, 111–113.

Newmark, P. (1988). Sober side up at giant AIDS meeting held in Stockholm. *Nature, 333,* 585.

Nzilambi, N., DeCock, K., Forthal, D., et al. (1988). The prevalence of infection with human immunodeficiency virus over a 10-year period in rural Zaire. *The New England Journal of Medicine, 318,* 276–279.

Odehouri, K., De Cock, K. M., Krebs, J.W., et al. (1989). HIV-1 and HIV-2 infection associated with AIDS in Abidjan, Cote d'Ivoire. *AIDS, 3,* 509–512.

Okware, S.I. (1987). Towards a national AIDS-control program in Uganda. *The Western Journal of Medicine, 147* (6), 726–728.

Osborne, J.E., Jensen, A.R., Cooke, M., et al. (1986). AIDS: Science, ethics and policy. *Issues in Science and Technology,* Winter, 40–54.

Osborne, J.E. (1986). An overview of the AIDS epidemic. *Emirates Medical Journal, 4,* 65–70.

Padian, N., Marquis, L., Francis, D.P., et al. (1987). Male-to-female transmission of human immunodeficiency virus. *Journal of the American Medical Association, 258,* 788–790.

Pape, J., & Johnson, W. (1989). HIV-1 infection and AIDS in Haiti. In R. Kaslow & D. Francis (Eds.) *The epidemiology of AIDS* (pp. 221–230). New York: Oxford University Press.

Parker, S.W., Stewart, A.J., Wren, M.N., Gollow, M.M., & Straton, J. (1983). Circumcision and sexually transmitted disease. *Medical Journal of Australia, 2,* 288–290.

Perlez, J., (1989, November 24). Zimbabwe resisting facts in AIDS epidemic. *The New York Times*, p. 7.

Peterman, T.A., Drotman, D.P., & Curran, J.W. (1985). Epidemiology of the acquired immunodeficiency syndrome (AIDS). *Epidemiologic Reviews, 7*, 1–21.

Piot, P., & Colebunders, R. (1987). Clinical manifestations and the natural history of HIV infection in adults. *The Western Medical Journal, 147*, 709–712.

Piot, P., & Mann, J.M. (1987). Bidirectional heterosexual transmission of human immunodeficiency virus (HIV). In J.C. Gluckman & E. Vilmer (Eds.), *Acquired immunodeficiency syndrome* (pp. 149–156). Amsterdam: Elsevier.

Piot, P., Plummer, F.A., Mhalu, F.S., et al. (1988). AIDS: An international perspective. *Science, 239*, 573–579.

Piot, P., Plummer, F.A., Rey, M.A., et al. (1987). Retrospective sero-epidemiology of AIDS virus infection in Nairobi prostitutes. *Journal of Infectious Disease, 155*, 1108–1112.

Piot, P., Taelman, H., Minlangu, K.B., et al. (1984). Acquired immunodeficiency syndrome in a heterosexual population in Zaire. *The Lancet, 2*, 65–69.

Pokrovsky, V.V., Yankina, Z.K., & Pokrovsky, V.I. (1987). Epidemiological investigation of the first case of the acquired immunodeficiency syndrome (AIDS) detected in the USSR. *Zhurnal Mikrobiologii, Epidemiologii I Immunobiologi (Moskva), 12*, 8–11.

Prewitt, K. (1988). AIDS in Africa: The triple disaster. In N. Miller & R.C. Rockwell (Eds.), *AIDS in Africa: The social and policy impact* (pp. ix–xii). Lewiston, NY: E. Mellen Press.

Pyle, G.F. (1986). *The diffusion of influenza: Patterns and paradigms.* Totowa, NJ: Rowman and Littlefield.

Quesenberry, C.P., Jr., et al. (1989). A survival analysis of hospitalization among patients with acquired immunodeficiency syndrome. *American Journal of Public Health, 79*, 1643–1647.

Quinn, T.C., & Mann, J. (1989). HIV-1 infection and AIDs in Africa. In R.A. Kaslow & D.P. Francis (Eds.), *The epidemiology of AIDS: Expression, occurrence, and control of human immunodeficiency virus type 1 infection* (pp. 194–220). New York: Oxford University Press.

Quinn, T.C., Zacarias, F.R., & St. John, R. (1989). AIDS in the Americas. *The New England Journal of Medicine, 320*, 1005–1007.

Quinn, T.C., Mann, J.M., Curran, J.W., & Piot, P. (1986). AIDS in Africa: An epidemiologic paradigm. *Science, 234*, 955–963.

Rhame, F., & Maki, D. (1989). The case for wider use of testing for HIV infection. The New England Journal of Medicine, 320, 1248–124.

Rich, V. (1987). AIDs arrives in the Soviet Union—(Official). *Nature, 326*, 3.

Ross, M.W. (1984). Predictors of partner numbers in homosexual men: Psychosocial factors in four societies. *Sexually Transmitted Diseases*, July–September, 119–122.

Rothenberg, R.B. (1983). The Geography of gonorrhea. *American Journal of Epidemiology, 117*, 688–694.

Ruef, C., Dickey, P., Schable, C.A., et al. (1989). A second case of the acquired immunodeficiency syndrome due to human immunodeficiency virus type 2 in the United States: The clinical implications. *The American Journal of Medicine, 86*, 709–712.

Sabatier, R.C. (1987). Social, cultural and demographic aspects of AIDS. *The Western Journal of Medicine, 147* (6), 713–715.

Safai, B., Johnson, K.G., et al. (1985). The natural history of Kaposi's sarcoma in the acquired immunodeficiency syndrome. *Annals of Internal Medicine, 103*, 744–750.

Salahuddin, S.Z., Markham, P.D., Popovic, M., et al. (1985). Isolation of infectious human t-cell leukemia/lymphotropic virus type III (HTLV-III) from patients with acquired immunodeficiency syndrome (AIDS) or AIDS-related complex (ARC) and from healthy carriers: A study of risk groups and tissue sources. *Proceedings of the National Academy of Science of the United States of America, 82*, 5530–5534.

Schmidt, N. (1988). Resources on the social impact of AIDS in Africa. In N. Miller & R.C. Rockwell (Eds.), *AIDS in Africa* (pp. 239–243). Lewisburg, NY: E. Mellen Press.

Schoepf, B.G., Nkera, R., Schoepf, C., et al. (1988). AIDS and society in Central Africa: A view from Zaire. In N. Miller & R.C. Rockwell (Eds.), *AIDS in Africa: The social and policy impact* (pp. 211–235). Lewiston, NY: E. Mellen Press.

Scitovsky, A.A. (1989a). Past lessons and future directions: The economics of health services delivery for HIV- related illnesses. In *New Perspectives on HIV-related illnesses: Progress in health services research* (pp. 7–12). National Center for Health Services Research and Health Care Technology Assessment, Public Health Service, U.S. Department of Health and Human Services.

Scitovsky, A.A. (1989b). The cost of AIDS: An agenda for research. In *New Perspectives on HIV-related illnesses: Progress in health services research* (pp. 197–208). National Center for Health Services Research and Health Care Technology Assessment, Public Health Service, U.S. Department of Health and Human Services.

Seale, J. (1988). Origins of AIDS viruses, HIV-1 and HIV-2: Fact or fiction? *Journal of the Royal Society of Medicine, 81*, 537–539.

Seale, J. (1989). Crossing the species barrier—Viruses and the origins of AIDS in perspective. *Journal of the Royal Society of Medicine, 82*, 519–523.

Shannon, G.W., & Pyle, G.F. (1989). The origin and diffusion of AIDS: A view from medical geography. *Annals of the Association of American Geographers, 79*, 1–24.

Shannon, G.W., & Spurlock, C. (1976). Urban ecological containers, environmental risk cells, and the use of medical services. *Economic Geography, 52*, 136–146.

Simonsen, J.N., Cameron, D.W., Michael, N.G., et al. (1988). Human Immunodeficiency virus infection among men with sexually transmitted diseases: Experience from a center in Africa. *The New England Journal of Medicine, 319*, 274–278.

Smith, E.W. (1989). Introduction. In *New Perspectives on HIV-related illnesses: Progress in health services research* (pp. 5–6). National Center for Health Services Research and Health Care Technology Assessment, Public Health Service, U.S. Department of Health and Human Services.

Smith, H. (1983). "AIDS: The Haitian connection." *MD, 27* (12), 46–52.

Spitzer, P., & Weiner, N., (1989). Transmission of HIV infection from a woman to a man by oral sex. *New England Journal of Medicine, 320:* 21.

Stein, B.S., Gowda, S., Lifson, J., et al. (1987). pH independent HIV entry into CD4-positive T-cells via virus envelope fusion to plasma membrane. *Cell, 49*, 659–668.

Sunderham, G., McDonald, R.J., et al. (1986). Tuberculosis as a manifestation of the acquired immunodeficiency syndrome (AIDS). *Journal of the American Medical Association, 256*, 362–366.

Taravella, S. (1988). States continue to seek approval for home care. *Modern Healthcare, 18*, 90.

Taylor, P.K., & Rodin, P. (1975). Herpes genitalis and circumcision. *British Journal of Venereal Diseases, 51*, 274–277.

The Institute of Cancer Research. (1984). The epidemiology of AIDS in Europe. *European Journal of Cancer and Clinical Oncology, 20*, 157–164.

Torrey, B.B., Way, P.O., & Rowe, P.M. (1988). Epidemiology of HIV and AIDS in Africa: Emerging issues and social implications. In N. Miller & R. Rockwell (Eds.), *AIDS in Africa: The social and policy impact* (pp. 31–54). Lewiston, NY: E. Mellen Press.

Travis, P., & Black, F.T. (1987). Heterosexuals importing HIV from Africa. *The Lancet, 1*, 325.

Turnock, B., & Kelly, C. (1989). Mandatory premarital testing for human immunodeficiency virus. The Illinois experience. *Journal of the American Medical Association, 261*, 3415–3418.

Vanbenbvoucke, C.M., & Verhoef, J. (1984). Acquired immune deficiency syndrome in the Netherlands. *European Journal of Clinical Microbiology, 3*, 62.

Van De Perre, P., Lepage, P., Kestelyn, P., et al. (1984). Acquired immunodeficiency syndrome in Rwanda. *The Lancet, 2,* 52–65.

Van Griensven, G., et al. (1988). Impact of HIV antibody testing on changes in sexual behavior among homosexual men in the Netherlands. *American Journal of Public Health, 78,* 175–1577.

Versi, A. (1980). Africa and the AIDS myth. *New African,* April, 9–12.

Vogt, M.W., Witt, D.J., & Craven, D.E. (1986). Isolation of cervical secretions of women at risk of AIDS. *The Lancet, i,* 525–527.

Von Reyn, C.F., & Mann, J.M. (1987). Global epidemiology of HIv and AIDS. *Western Journal of Medicine, 147,* 694–701.

Waite, G. (1988). The politics of disease: The AIDS virus in Africa. In N. Miller & R. Rockwell (Eds.), *AIDS in Africa: The social and policy impact* (pp. 145–164). Lewiston, NY: E. Mellen Press.

Weber, J.N., Carmichael, D., et al. (1984). Clinical aspects of the acquired immune deficiency syndrome in the United Kingdom. *British Journal of Venereal Disease, 60,* 253–257.

Weber, J.N., & Weiss, R.A. (1988). HIV infection: The cellular picture. *Scientific American, 259,* 101–109.

Weiss, R., & Thier, S. (1988). HIV testing is the answer—What is the question? *The New England Journal of Medicine, 319,* 1010–1012.

Weiss, S.H., & Biggar, R.J. (1986). The epidemiology of human retrovirus–associated illnesses. *Mount Sinai Journal of Medicine, 53,* 579–591.

Wendler, J., Schneider, J., Garas, B., et al. (1986). Seroepidemiology of human immunodeficiency virus in Africa. *British Medical Journal, 293,* 782–785.

Wendt, D., Sadowski, L., Markowitz, N., et al. (1987). Prevalence of serum antibody to human immunodeficiency virus among hospitalized intravenous drug abusers in a low-risk geographic area. *The Journal of Infectious Diseases, 15,* 151–12.

Willis, D.P. (1989). The social consequences of AIDS. In *New Perspectives on HIV-related illnesses: Progress in health services research* (pp. 79–83). National Center for Health Services Research and Health Care Technology Assessment, Public Health Service, U.S. Department of Health and Human Services.

Winkelstein, W., Lyman, D.M., et al. (1987). Sexual practices and risk of infection by the human immunodeficiency virus. *Journal of the American Medical Association, 257,* 321–325.

Winslow, C.E.A. (1980). *The conquest of epidemic disease.* Madison: University of Wisconsin Press.

Wkofsy, C.B., Cohen, J.B., & Hauer, L.B. (1986). Isolation of AIDS-associated retrovirus from genital secretions of women with antibodies to the virus. *The Lancet, i,* 527–529.

Wolfheim, J.H. (1983). *Primates of the world.* Seattle: University of Washington Press.

Wood, W.B. (1988). AIDS north and south: Diffusion patterns of a global epidemic and a research agenda for geographers. *The Professional Geographer, 40,* 266–279.

World Health Organization. (1986). Acquired immunodeficiency syndrome (AIDS). *Weekly Epidemiological Record, 61,* 69–76.

World Health Organization. (1987a). Acquired immunodeficiency syndrome (AIDS): Situation in the WHO European region as of 31 December 1986. *Weekly Epidemiological Record, 62,* 117–124.

World Health Organization. (1987b). Acquired immunodeficiency syndrome (AIDS): Situation in the WHO European region as of 31 March 1987. *Weekly Epidemiological Record, 62,* 229–232.

World Health Organization. (1987c). Acquired immunodeficiency syndrome (AIDS). *Weekly Epidemiological Record, 62:* 221–223.

World Health Organization. (1989a). Acquired immunodeficiency syndrome (AIDS): Global projections of HIV/AIDS. *Weekly Epidemiological Record, 64,* 229–231.

World Health Organization. (1989b). Acquired immunodeficiency syndrome (AIDS)—Data as of 30 November 1989. *Weekly Epidemiological Record, 64,* 369–370.

World Health Organization. (1989c). Acquired immunodeficiency syndrome (AIDS): HIV seroprevalence survey. *Weekly Epidemiological Record, 64,* 197–204.

World Health Organization. (1990). Acquired immunodeficiency syndrome (AIDS)—Data as of 28 February 1990. *Weekly Epidemiological Record, 65,* 61–62.

Wright, K., (1990). Mycoplasmas in the AIDs spotlight. *Science, 248:* 682–683.

WuDunn, S. (1990, March 30). Outbreak of AIDs among drug users alarms China. *The New York Times,* p. A8.

Yeager, R. (1988). Historical and ecological ramifications for AIDS in Eastern and Central Africa. In N. Miller & R. Rockwell (Eds.), *AIDS in Africa: The social and policy impact* (pp. 71–81). Lewiston, NY: E. Mellen Press.

Zuckerman, A. (1986). AIDS and insects. *British Medical Journal, 292,* 1094–1095.

Index

Acquired immune deficiency
syndrome (AIDS)
behavior patterns in spread of,
57, 61, 158-164
costs of care for, 142-143,
144-147
Delphi projections of, 4,5
epicenter of, 50, 64, 66, 68, 73
incubation period of, 140
patterns of infection in,
37-40, 155-157
prevention of, 33, 35, 138-139
rank of, in years of lives lost
(U.S.), 134
reported cases of, 1-2; *see
also* various countries
in rural vs. urban populations,
69-74, 81, 95
social manifestations of,
139-140
WHO estimates of, 3
Africa, 61, 59-89, 164; *see also*
Central Africa, North Africa,
West Africa
AIDS belt in, 66, 72, 73
ethnic variation in, 66

HIV infection rates in, 6, 50,
59-60, 66
HIV-2 in, 19-20, 49
long-distance truck drivers in,
78-79, 164
matrilineal belt of, 61
medical testing in, 48, 62
migration from, 54, 72, 76, 92,
165
monkeys of, 20, 32, 51, 55-56,
81-83
and origin of AIDS, 48-56
regional distribution of AIDS
in, 64-65
sub-Saharan, 10, 11, 37, 48, 56
59, 66, 155
transportation routes in, 54,
70, 79, 80
tribal distribution of AIDS in,
81
AIDS quotients, in U.S. states,
114-117
AIDS-related associated virus
(ARV), 14
AIDS-related complex (ARC),
15